PUBLIC HEALTH IN THE 21ST CENTURY

# OCCUPATIONAL HEALTH AND SAFETY IN THE CHEMICAL AND BIOLOGICAL LABORATORY HANDBOOK

# PUBLIC HEALTH IN THE 21ST CENTURY

Additional books and e-books in this series can be found on Nova's
website under the Series tab.

PUBLIC HEALTH IN THE 21ST CENTURY

# OCCUPATIONAL HEALTH AND SAFETY IN THE CHEMICAL AND BIOLOGICAL LABORATORY HANDBOOK

MARIA PIA GATTO, PHD

Copyright © 2019 by Nova Science Publishers, Inc.

**All rights reserved.** No part of this book may be reproduced, stored in a retrieval system or transmitted in any form or by any means: electronic, electrostatic, magnetic, tape, mechanical photocopying, recording or otherwise without the written permission of the Publisher.

We have partnered with Copyright Clearance Center to make it easy for you to obtain permissions to reuse content from this publication. Simply navigate to this publication's page on Nova's website and locate the "Get Permission" button below the title description. This button is linked directly to the title's permission page on copyright.com. Alternatively, you can visit copyright.com and search by title, ISBN, or ISSN.

For further questions about using the service on copyright.com, please contact:
Copyright Clearance Center
Phone: +1-(978) 750-8400          Fax: +1-(978) 750-4470          E-mail: info@copyright.com.

## NOTICE TO THE READER

The Publisher has taken reasonable care in the preparation of this book, but makes no expressed or implied warranty of any kind and assumes no responsibility for any errors or omissions. No liability is assumed for incidental or consequential damages in connection with or arising out of information contained in this book. The Publisher shall not be liable for any special, consequential, or exemplary damages resulting, in whole or in part, from the readers' use of, or reliance upon, this material. Any parts of this book based on government reports are so indicated and copyright is claimed for those parts to the extent applicable to compilations of such works.

Independent verification should be sought for any data, advice or recommendations contained in this book. In addition, no responsibility is assumed by the Publisher for any injury and/or damage to persons or property arising from any methods, products, instructions, ideas or otherwise contained in this publication.

This publication is designed to provide accurate and authoritative information with regard to the subject matter covered herein. It is sold with the clear understanding that the Publisher is not engaged in rendering legal or any other professional services. If legal or any other expert assistance is required, the services of a competent person should be sought. FROM A DECLARATION OF PARTICIPANTS JOINTLY ADOPTED BY A COMMITTEE OF THE AMERICAN BAR ASSOCIATION AND A COMMITTEE OF PUBLISHERS.

Additional color graphics may be available in the e-book version of this book.

## Library of Congress Cataloging-in-Publication Data

ISBN: 978-1-53615-526-6

*Published by Nova Science Publishers, Inc. † New York*

*This book is dedicated to my mother*

# CONTENTS

| | | |
|---|---|---|
| **Preface** | | **ix** |
| **Chapter 1** | Introduction | **1** |
| **Chapter 2** | The Analytical Laboratory | **17** |
| **Chapter 3** | Chemical Risk | **43** |
| **Chapter 4** | Biological Risk | **69** |
| **Chapter 5** | Radiological Safety | **89** |
| **Chapter 6** | Collective Protection Equipment | **107** |
| **Chapter 7** | Personal Protection Equipment | **119** |
| **Chapter 8** | Laboratory Fire Safety | **141** |
| **Chapter 9** | First Aid | **153** |
| **Chapter 10** | Laboratory Safety Standards | **169** |
| **Chapter 11** | Glossary | **179** |
| **Author's Contact Information** | | **205** |
| **Index** | | **207** |
| **Related Nova Publications** | | **213** |

# PREFACE

Occupational health and safety refers to the whole series of prevention measures and protection, technical measures, organizational solutions, and procedures, that must be adopted by the employer to avoid dangerous situations for their employees. While it is true that total security exists in the absence of dangers, and this is a difficult concept to translate into real life, in the absolute sense, it is also true that the application of safety rules makes the occurrence of adverse events and incidents more difficult and it always results in a better quality of life. Each analytical laboratory should provide its own "Good Laboratory Practices", procedures must be the result of the experience and must be known, correctly interpreted, shared, and respected by all the staff of the laboratory.

The idea of creating a manual on occupational health risks in a chemical and biological analysis laboratory derives from the need to provide technical support for the various figures involved in activities requiring the safe use of chemical and biological agents in lab analysis. The purpose of this work is to provide workers with more detailed information on the main possible causes of risk in an analysis laboratory, as well as on the most appropriate means of individual and collective protection to be used at work. The manual also gives an overview of main laboratory safety standards pertain to clothing and equipment as well as procedures and lab design.

This manual can be a valuable approach to knowledge of laboratory risks in order to eliminate or at least minimize them. Finally, the handbook can also be a support to the drafting of internal safety standards, both for general and specific procedures.

*Chapter 1*

# INTRODUCTION

Staff working in the analytical laboratories are exposed to frequently underestimated risks by those who are directly involved, who very often become aware of the existence of a specific hazard only when serious accidents occur.

In the laboratories, in fact, the risk is usually invisible and, consequently, more dangerous than other work activities [1].

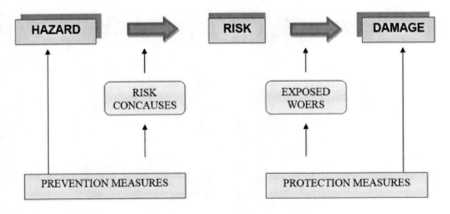

Figure 1. Risk management scheme.

All risk factors contributing to a particular activity must be identified in order to define standard procedures to limit the risk as much as possible. In these procedures, the general measures and principles for the prevention of risks and the specific protection and prevention measures adopted must be indicated [2].

The head of laboratory is obliged to implement all possible activities for the correct protection and risks measures to safeguard the personnel exposed to the damages that may result from the execution of the work, according to Figure 1.

## Table 1. Definitions for case histories[*]

| Term | Definition |
| --- | --- |
| Accident | An undesired and unplanned (but not necessarily unexpected) occurrence of a sequence of events that results in (at least) a specified level of loss (injury, death, or property damage). |
| Hazard | A chemical or physical condition set of conditions of a system that, together with other conditions in the environment of the system will lead inevitably to an accident. |
| Incident | The loss of containment of material or energy; not all events propagate into incidents; not all incidents propagate into accidents. |
| Consequence | A measure of the expected effects of the results of an incident. |
| Likelihood | A measure of the expected probability or frequency of occurrence of an event. This may be expressed as a frequency, a probability of occurrence during some time interval, or a conditional probability |
| Risk | A measure of human injury, environmental damage, or economic loss in terms of both the incident likelihood and the magnitude of the loss or injury. |
| Risk analysis | The development of a quantitative estimate of risk based on an engineering evaluation and mathematical techniques for combining estimates of incident consequences and frequencies |
| Risk assessment | The process by which the results of a risk analysis are used to make decisions, either through a relative ranking of risk reduction strategies or through comparison with risk targets. |
| Scenario | A description of the events that result in an accident or incident. The description should contain information relevant to defining the root causes. |

[*] Center for Chemical Process Safety (CCPS), Guidelines for Consequence Analysis.

## 1.1. RISK ASSESSMENT PROCEDURES

The risk is defined as the probability that a substance, object or situation may create danger to the operator under specific conditions. It is the combination of two factors:
- the likelihood of an accident or adverse event occurring; i.e., the probability of an unsafe event or condition occurring.
- the seriousness of the damage related to the occurrence of the adverse event; i.e., the possible effect of the unsafe condition in a predictably worst situation.

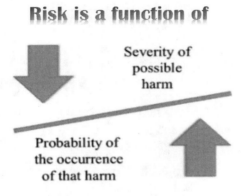

Figure 2. The function of risk assessment.

Risk assessment is a systematic process of assessing potential risks that may be involved in a projected activity or enterprice[3]. It requires, for each environment considered or workplace, a series of successive and consequent operations:

- the recognition of the sources of risk present in the working cycle;
- the identification of the consequent potential risks of exposure with regard to the operation of the functions;
- estimating the amount of exposure risks associated with the identified preventive interest situations.

**Table 2. Classification of major risks in the analytical laboratory**

| General risks | Environmental | Microclimate<br>Lighting<br>Noise<br>Overcrowding |
|---|---|---|
| Specific risks | Physical | Mechanical<br>Biochemical<br>Electrical<br>Thermal<br>Exposure to video terminals |
| | Chemical | |
| | Biological | |
| | Ionizing/non-ionizing radiation | |
| | Psychosocial | |

The risk assessment must take into account a number of parameters which cannot be neglected, such as:

- the intrinsic hazard of each substance handled
- the chemical-physical characteristics (state of aggregation, vapor pressure, boiling point, particle size, etc.) of the reagents used
- the temperature at which the tests are conducted
- concentration and density of reagents adopted
- the absorption pathways
- the capacity of the hazardous substances treated to penetrate the body through the different absorption pathways, also in relation to their state of aggregation and, if they are in the solid state, if they are in compact masses or in flakes or in dusty form, or if they are contained in a solid, non-dusty matrix that reduces or limits dispersion
- the quantities of hazardous substances used (global and for each worker)

# Introduction 5

- storages
- transport deposits
- the processes adopted and the equipment involved
- times and frequency of use
- the procedures for the possible neutralisation and disposal of waste
- electrical safety
- fire prevention measures
- organization of work
- handling of loads
- the professionalism of the operators
- information and training of operators
- measures the prevention measures involved
- measures the collective and individual safety measures present at the workplace
- equipment and personal protective devices to be used in the event of an emergency
- the presence of a procedure to be followed in the event of an emergency
- control and monitoring systems of the safety levels set
- occupational exposure limit values or biological limit values
- conclusions drawn from any health surveillance actions already undertaken

The risk assessment could lead to the identification of laboratory sectors, homogeneous groups of workers or individual workers, and therefore, consequently, with different actions to be undertaken.

In particular, it is essential to identify:

1. No exposure risk;
2. The presence of a controlled exposure within the limits of acceptability required by law;
3. The presence of a risk of exposure.

In the first case there will be no problems associated with the elaboration of the work; in the second, the situation was kept under periodic control; in the third case the necessary preventive and protective measures must be implemented (Figure 3).

The results of the evaluation are given in a document specifying the criteria adopted for evaluation, the prevention and protection measures applied and/or the programme for their implementation and, of course, the identification of exposed workers. The document must also contain the results of the environmental monitoring of hazardous chemical agents and the data of biological monitoring carried out anonymously on workers (if necessary) [4].

Obviously, the process of recognizing the danger and, as a result, the risk assessment are not simple actions. A risk matrix chart can help you visualize the concept of risk and provide a scoring system to understand the level of risk involved in all daily activities.

**Table 3. Risk matrix**

| Risk probability | Risk severity | | | | |
|---|---|---|---|---|---|
| | Cathastrophic A | Hazardous B | Major C | Minor D | Negligible E |
| Frequent 5 | 5A | 5B | 5C | 5D | 5E |
| Occasional 4 | 4A | 4B | 4C | 4D | 4E |
| Remote 3 | 3A | 3B | 3C | 3D | 3E |
| Improbable 2 | 2A | 2B | 2C | 2D | 2E |
| Extremely improbable 1 | 1A | 1B | 1C | 1D | 1E |

# Introduction

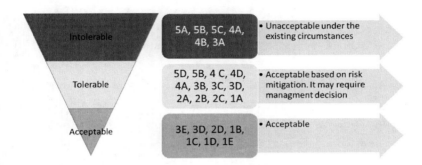

Figure 3. Criteria for classification of risk.

If the risk score is at any point in the red, it would not be prudent to continue until further measures and controls can be in place to reduce the severity and probability to acceptable levels.

If the risk score is in the yellow, you can continue with the moderate risk, but additional checks and monitoring would be advisable. If conditions worsen, or controls are not effective, the risk can return to the red zone. If the risk score is green, the risk level is acceptable and you can proceed as planned. This does not mean that something bad will neves happen, because we can never eliminate bad outcomes, but, if you go south, the consequences would most likely be acceptable. Potential gains outweigh the risk involved and the existing controls and mitigations are sufficient to maintain the manageable risk.

## 1.2. GENERIC RISKS IN THE ANALYTICAL LAB

### 1.2.1. Microclimate

The biological or chemical laboratory must have suitable equipment for the control of the microclimate. Simple devices such as the classic wall thermometer and recording systems are required. The microclimate must be taken into account in case of discomfort, e.g., lack of heating, humid environment, abnormal operation of the air conditioning system. As for the temperature, it should normally be kept around 18°C, relative humidity within approximately 50±5 percentage, ventilation maintained at 0.1-0.2 m/s.

The measuring instruments of temperature and humidity must be placed in a central position of the environment which is influenced by changes in doors and/or windows, sunlight and radiators as little as possible, and near the maximum concentration of people. In very large environments, it is advisable to provide surveys in different places. It is advisable to prepare an operating procedure with the name of the measuring instruments for maintenance and the periodicity of the measures. However, if the laboratory does not have automatic recording systems, the temperature and other parameters must be measured and the values given manually at predeterminated intervals on appropriate modules. The speed and the ventilation directions must also be determinated following a predetermined program, using anemometers, equipped with a probe suitable for the measuring range.

In the laboratory the possible heat-related risks have two causes: high temperatures (when using heat sources) and low temperatures (manipulation of cryogenic compounds). Free flames are commonly used in microbiology for the sterilization of loops and needles or for the flambation of containers. In addition, incubators, dry stoves, flask furnaces and autoclaves are readily available in any laboratory and may cause burns. Cryogenic liquids and gases, such as dry ice or liquid nitrogen can also cause burns. The cryogenic liquid is defined as a liquid with a normal boiling point below −150°C.

Cryogenic fluids must be handled, stored, and used only in containers or systems designed in accordance with safety standards, procedures or practices.

- To withstand extreme low temperatures, all components, including piping, valves etc., must consists of appropriate materials that last at low temperatures.
- To avoid overpressure, the pressure discharge systems must be used in the piping.

# Introduction 9

- Cryogenic containers or any part of these systems that could be deactivated must be equipped with an overpressure valve. These valves must be positioned to face forward.
- All necessary PPE must be worn durinfg transfer operations, including the cryogenic containers opening. People have to move slowly during the transfer. More attention must be paid with non-insulated tubes and other components.
- Open transfers are only permitted in well-ventilated areas.
- Never use a funnel when transferring cryogenic liquids.
- Pliers or other similar devices should be used to immerse and remove objects from cryogenic liquids.
- Hazard reviews are necessary on all newly purchased instruments, constructed or modified using cryogenic materials.

## 1.2.2. Noise

Noise is an important problem for health protection in the workplace. In laboratories, it is difficult to find instruments or other equipment that produces noise levels that cause hearing damage to the worker and that should be soundproof. Sometimes, however, the noise level of centrifuges, mechanical agitators, or automatic instruments can generate a daylong background noise that cannot only constitute physical damage to the workers, but also causes distractions. In order to eliminate or reduce irritating noises, the friction points between the instruments and support surfaces must be reduced with the installation of soundproof panels.

In laboratory environments, it is not possible to establish very precise values for the levels of noise acceptability. In any case, it should vary from 45 to 55 db, and never exceed 85 db. The sound level gauge measures noise, reacting at certain frequencies in a similar way to the human ear. The sound waves perceived by the microphone are transmitted to analyzer wire, through a special amplifier and sound measurement must take into account the reflex phenomenon of the sound waves (against the objects, against walls, against the operator himself) that causes interference effects.

### 1.2.3. Lighting

Good lighting, natural or artificial, is necessary in the working environment in order to allow a high degree of visual efficiency, thus reducing fatigue and the possibility of errors and accidents. In fact, it is known that the visualization capacity varies according to the intensity of the light. The lighting characteristics must take into account both the quantity and the quality of the light, with particular attention to the color and the reduction of the dazzling phenomena.

The intensity of the light and the number and position of the lighting fixtures must be associated with the type of work carried out, as well as with the age of the persons performing the operations. The unit of measurement of light intensity, the lux (symbol lx) is the unit derived from the illuminance and luminous emissions, measuring the luminous flux by unit area. It is equal to one lumen per area (expressed in $m^2$). The lux meter is the instrument for measuring illumination in workplaces and must be able to reconstruct exactly what would be the of the human eye's response by means of special optical filters. A photodiode converts the incident light into an electrical signal that is then displayed.

The lux (symbol lx) is the SI unit of measurement of the light intensity and luminous emittance, measuring luminous flux per unit area. It is equal to one lumen per square metre. In photometry, this is used as a measure of the intensity, as perceived by the human eye, of light that hits or passes through a surface.

According the DIN 5035-2, the recommended artificial illuminance values for the lighting parameters for indoor and outdoor workspaces, the luminous intensity suggested for the offices is 300-750 lux, depending on the different types of work [5]. In the workplaces, an intensity of 250-500 lux for the natural source must be maintained, and 100-300 lux for the artificial ones. Lastly, with regard to the quality of artificial light, fluorescent lamps emitting yellow radiation must be privileged, ensuring maximum visual acuity and reducing the symptoms of fatigue, instead of those emitting radiation blue-green.

## 1.2.4. Overcrowding/Working Alone

A working area of at least 2 m$^2$ must be assigned to each individual worker. The rooms must have a height of not less than 3 m and a cubic height of 10 m$^3$. Finally, cozy and spacious rooms for users, with waiting room and bathrooms are needed. On the other hand, even working without another suitably qualified person present, and especially outside of normal hours in laboratory, is a potentially dangerous practice and should not be undertaken. Therefore, working alone outside of normal hours in laboratories is even more potentially hazardous, and is positively discouraged.

## 1.2.5. Risks Due to Mechanical Factors

Mechanical traumas can be caused by slipping on wet or dirty floors, heavy objects falling, breaking of the glassware, and use of sharp or pointed objects and use of sharp-edged containers.

To avoid this type of accident it is good practice to place the heaviest material at a reduced height, to make it less complicated for its staff, and, next to this, furnish the laboratory with shelves and shelves with raised edge. The operator must wear appropriate safety gloves when handling sharp objects and materials; broken glass must be disposed of and not reused. Sometimes the incident may occur due to collisions with objects deposited in corridors and transit areas. The danger of violent impact against these elements increases in an emergency. Therefore it is necessary to pay close attention to the arrangement of objects along the corridors, including fire extinguishers, water dispensers, waste bins, positioning them in places where they do not constitute obstacles to movement or passage. When manipulating sharp objects and materials, the operator should wear appropriate safety gloves. Broken glassware must be disposed of and not re-used. Sometimes the incident can occur due to collisions with objects deposited in corridors and transit areas. The danger of violent impact against these elements increases in the event of an emergency. Therefore, it is necessary to pay close attention to the arrangement of objects along the corridors, including fire extinguishers, water dispensers, waste bins, placing them in places where they do not constitute obstacles to movement or passage. Additional hazards may occour during the use of centrifuges, mechanical agitators, and vending machines, for contact between the moving parts of the equipment with the fingers, hair and apparel of the operator. Mechanical accidents can also be caused by the sudden release of energy by pressurized equipment, such as the autoclave (release of steam) and cylinders (release of compressed gas). The sudden opening of a containment system, due to breakage or incorrect maneuvers during use, can throw objects at a distance with particular violence, which the operator can invest. The cylinders must be held upright, anchored to the walls to avoid possible falls or shocks. They must not be kept close to flammable materials and must be fitted with a safety valve. The gas pipes must be checked periodically and in any case before each use. Combustible gases and comburants must be housed in suitable external constructions, at a safe distance from the laboratory, and never stored together.

### 1.2.6. Biomechanical Factors

The identifiable dangers for the health of the operators are above all the posture assumed during the work activities, mainly in front of the video terminals, and the handling of the loads. In the laboratory, the wrong posture represents a real and widespread danger, while the handling of loads is not particularly onicious. The ergonomic structure of the worker must take into account a number of environmental and position variables, and in particular of time. In fact, the damage caused to the health of the operator is due to incorrect and prolonged postural attitudes. The neck, back, hands, arms and shoulders generally support musculoskeletal overload disorders. Usually the disorders that cause pain and discomfort are reversible and originate from static positions kept too long and from poorly structured jobs.

## 1.2.7. Dangers from Electricity

Almost all laboratory instruments are electrically powered. Therefore, electrocution is a very widespread cause of accidents, both for the personnel who routinely use these equipments for the work and for the operators involved in their maintenance [6]. The main dangers associated with electricity are electrical shock and fire. The electric shock occurs when the body becomes part of the electric circuit, both when an individual comes into contact with both wires of an electrical circuit, a wire of an energized circuit and the ground, or a metallic part that has become energized by contact with an electrical conductor. The severity and effects of an electrical shock depend on a number of factors such as the pathway through the body, the amount of current, the length of time of the exposure, and whether the skin is wet or dry. Water is a great conductor of electricity, allowing current to flow more easily in wet conditions and through wet skin. The effect of the shock can vary from a mild formicle to severe burns to cardiac arrest [7]. Only trained electricians must perform repairs of high-voltage or high-current equipment [8]. However, if urgent intervention by the laboratory staff is necessary, it is mandatory to isolate the electrical circuit before proceeding with the repair, avoiding, however, to approach with the hands or other parts of the body wetted or in contact with the water. In case of fire, it could spread through the wires of the electrical system in other areas of the laboratory. It is therefore necessary to observe special precautions regarding the electrical equipment subjected to heating, such as the resistance contained in dry stoves, sterilizers and mittles. This material must be protected with insulation, checked and periodically cleaned of any flammable residues. The power sockets and connection cables must also be inspected at least once a year to verify the suitability of the contacts and to identify any deterioration points.

## 1.2.8. Video Terminals

When producing images on the screen, the computer generates not only visible light, but also electromagnetic radiations of different wavelengths: X-rays, ultraviolet and infrared radiation, electromagnetic fields and radiofrequencies. The intensity of such radiations is so weak that it is hardly measurable or it is only within the reading limits of the instruments, out of the video, as has been shown by a series of measurements on instruments that differed in brand, model and condition of maintenance. A work involving a computer pose a health risk in relation to the duration of the exposure, the characteristics of the work, the hardware and software characteristics and the characteristics of the work environment.

However, video terminals are a serious danger for visual fatigue and musculoskeletal disorders due to improper posture, support and unsuitable seating. This last form of danger is often underestimated and ignored in the laboratory to the point that video terminals are often installed on the first available support, ignoring what the posture taken by those who will use them. In order to prevent and avoid disturbances, it is not sufficient to pay attention to ergonomics in the planning of workplaces, but also the correct behavior of the operators is required. A suspension of the activity of 15 minutes is recommended every two hours of continuous work.

## 1.2.9. Psychological Risks

The lack of adequate stimuli for the operator can lead to disinterestingness and little attention to work, which inevitably results in boredom and negative involvement of the rest of the staff. On the other hand, over-stimulation involves over-excitation, especially in those operators who fail to absorb and metabolize innovations. In fact in them comes an unconscious feeling of inadequacy that turns into hyperactivity and discontent, also involving the rest of the staff in negative.

Finally, whenever the working conditions are insufficient both in terms of quality and quantity, and in organizational terms, the phenomenon of

*Introduction* 15

discomfort is realized. This condition translates into various phases, linked to different levels of severity, depending on the time and intensity of exposure to discomfort.

Overcharging disorders are psychological or psychosomatic, such as headache, fatigue, irritability, nervous tension, anxiety, depression, insomnia, digestive problems. These symptoms are typical of stress syndrome and can occur when the type of work or the workloads distribution systematically exceeds the person's capacities, depending on attitudes or personality.

## REFERENCES

[1] Carson, P.A., Mumford, C.J. (2013) *Hazardous Chemicals Handbook Hazardous Chemicals Handbook*, pp. 1-378.

[2] Galante, E.B.F., Costa, D.M.B., França, T.C.C., Viaro, R.S. Risk assessment in a chemical laboratory (2016) *Occupational Safety and Hygiene IV - Selected, Extended and Revised Contributions from the International Symposium Occupational Safety and Hygiene*, 2016, pp. 105-109.

[3] ISOGuide 73:2009. *Risk management - vocabulary - guidelines for use in standards*. Geneva: International Standards Organisation.

[4] Angerer, J., Ewers, U., Wilhelm, M. Human biomonitoring: State of the art (2007) *International Journal of Hygiene and Environmental Health*, 210 (3-4), pp. 201-228.

[5] DIN 5035-2. *Artificial lighting recommended values for lighting parameters for indoor and outdoor workspaces.* 1990 Edition, September 1990, Deutsches Institut fur Normung E.V. (DIN).

[6] Florida State University EH&S (2016). *FSU Safety Manual - Electrical Safety in the Laboratory* (March 2015.)

[7] Zakharyuta, Anastasia and Şen, Canhan and Avaz, Merve Senem and Akkaş, Tuğçe and Pürçüklü, Sibel and Baytekin Birkan, Tuğba and Gönül, Turgay and Yerdelen, Bilge and Cebeci, Fevzi Çakmak and İnce,

(2016) Adnan Sabanci University, Istanbul. *Laboratory Safety Handbook*. ISBN 978-605-9178-58-7.

[8] Furr, A. K. (2000). *CRC Handbook of Laboratory Safety*. Care and Use of Electrical Systems (5th ed.) (pp. 328-336). United States of America: CRC Press LLC.

*Chapter 2*

# THE ANALYTICAL LABORATORY

## 2.1. THE BUILDING

In the risk assessment, the study of the floor plan of the building and the distribution of different areas is crucial. Typically, the analytical laboratory is located in a dedicated building. If it is embedded in a more complex structure, it is a good rule to locate it on the ground floor (preferably not buried), well protected against atmospheric agents and humidity, and with sufficient thermal insulation.

If the laboratory is located on the upper floors of the building, it is necessary to install sufficient safety scales (at least 1 for every 400 m$^2$ of surface and, in any case, never less than two). Each laboratory must have at least two exits and the safety pathways to reach the outside or "safe places" must be less than 15-20 m and well-marked. The laboratory area must be confined by fire doors. If several laboratories are present in the same building, it is a good idea to identify those who are most at risk.

The storage points of flammable chemicals, hazardous wastes from laboratory activities, toxic gases, compressed and dissolved gas cylinders, liquefied cryogenic gas tanks, technical premises such as the power plant, the local boiler, the compressor room, etc must be well identified [1].

18                                    *Maria Pia Gatto*

Careful analysis of fire-fighting facilities and compartments, including the verification of bearing structures, vertical and horizontal separators, fire-resistant doors, fire reaction of floors, walls and furnishings, staircases and elevetors (presence of ventilation chimneys, protected type, smoke-proof) must be carried out. The routes of exodus must be present and easily accessible and, in case of emergency, the ease of access to all the areas, both internal and external, must be ensured to the intervention of the rescue teams.

The Occupational Safety and Health Administration (OSHA) sets standards for the workplace to keep workers safe.

OSHA Safety Signs follow regulations to help prevent incidents and illnesses. Specific colors draw attention to different messages and also ensure the signs, legends, and text stand out.

## 2.2. SAFETY SIGNS

**Table 4. Safety sign colour codes**

| Colour | Indications |
| --- | --- |
| Prohibition | Signal that prohibits behavior that could cause it to run or cause a hazard |
| Mandatory | Signal prescribing specific behavior - obligation to wear explicit PPE (Personal Protective Equipment) |
| Warning | Signal that warns of a risk or danger |
| Safe Condition | Rescue and safe signals (Doors, exits, safety routes, protected positions) |

# The Analytical Laboratory

## Table 5. Prohibition signs (circular shape, white color, red frame)

| | | | | | |
|---|---|---|---|---|---|
| | Smoking strictly prohibited | | Prohibition of access to non-authorized persons | | No smoking or using free flames |
| | Do not drink water from the laboratory taps | | No eating or drinking in this area | | Do not use mobile phones |
| | Prohibited to turn off flame with water | | Do not touch the surfase | | Prohibited to delete waste in the exhausts |
| | Forbidden the pipetting with the mouth | | Prohibited to deposit materials | | Do not use the lift |

## Table 6. Fire fighting equipment (square shape, red color)

| | | | | | |
|---|---|---|---|---|---|
| | Fire alarm | | Fire alarm call point | | Telephone for fire-fighting interventions |
| | Emergency stop | | General electric switch to use only in case of fire | | Automatic closing door |
| | Fire exstinguisher | | Fire exstinguisher trolley | | Fire ladder |
| | Fire blanket | | Fire hosereel | | Fire Hydrant |

## Table 6. (Continued)

| | | | | | |
|---|---|---|---|---|---|
| | Direction to follow | | Interception valve | | Sprinkler |
| | Presence of smoke detectors | | In case of fire use stairway for exit | | Fire evacuation horn |

## Table 7. Mandatory signs (circular shape, light blue color)

| | | | | | |
|---|---|---|---|---|---|
| | Wear protective helmet | | Goggles must be worn | | Respirators must be worn |
| | Protective gloves must be worn | | Protective footwear must be worn | | Ear protectors must be worn in this area |
| | Wear faceshield | | Protective induments must be worn | | Wash your hand |
| | Throw waste in the proper containing box | | Obligation for carts to move slowly | | Hang in safety gas cylinders |
| | Disconnect the switch | | Grounding symbol | | Move slowly |

# The Analytical Laboratory

## Table 8. Safety signs (square shape, green color)

| | | | | | |
|---|---|---|---|---|---|
| | Assembly point | | Emergency telephone | | Emergency stairsway |
| | Emergency exit only | | Emergency exit | | Emergency stretcher |
| | Firstaid box | | Emergency shower | | Emergency eyewash |
| | Defibrillator | | Break glass to obtain access | | Emergency stop push button |

## Table 9. Warning signs (triangular shape, yellow or orange color)

| | | | | | |
|---|---|---|---|---|---|
| | Confined space | | Higly flammable | | Risk of explosion |
| | Oxidant agent | | Low temperature/freezing conditions | | High temperature |
| | Hot surface | | Emergency shower | | Radiation hazard |

## Table 9. (Continued)

| | | | | | |
|---|---|---|---|---|---|
| | High voltage | | Magnetic field | | Emergency stop push button |
| | Flammable gas | | Falling objects | | Crushing of hands |
| | Corrosion substance | | Poison | | Beware of opening door |
| | Mind your head | | Trip hazard | | Sharp elements |

## 2.3. THE INSTALLATIONS

An important phase of the risk assessment in the analylical laboratories is the identification of the installations.

### 2.3.1. Electric System

Electrical systems must be grounded and have high-sensitivity differential switches (lifesaving) or other equivalent protection systems.

For fire prevention, electrical installations:

- must not constitute a primary cause of fire or explosion;
- must not feed the propagation of fire;

# The Analytical Laboratory

- must be subdivided in such a way that the entire system is not removed from the service;
- must have control devices located in protected positions and must show clear indications of the circuits to which they refer.

The general electrical panel must be in an easily accessible position, signaled and protected from fire to allow the electrical system of the activity to extinguish. Refrigerators with flammable materials stored must have an explosion-proof electrical system.

## 2.3.2. Gas Distribution

Laboratory gases are treated with appropriate purification techniques and certified by analysis with validated methods, which guarantee their degree of purity. The products obtained are then compressed into cylinders of various size and capacities, and stored in warehouses for delivery. This system has the disadvantage of the need to equip the laboratory with distribution lines and deposits in compliance with the current safety regulations with the application of a series of accessories for the control of distribution and distribution. It is good practice to put on the pipes inside which the gases pass safety labels indicating the nature of the gas flowing through the pipe.

These problems have favored the development and diffusion of gas generators that can be connected directly to the device. They produce the amount of gas required for contingent consumption and do not require installations in accordance with safety standards. The generation and the direct supply ensures also the constancy of the degree of purity of the gas, without running the risk of pollution in the production chain, purification, filling of the cylinder (which can cause variability between the batches) transport and installation in the distribution line.

Gases can be flammable, corrosive and toxic. Depending on the degree of danger and harmfulness of the gas, the regulations in force for the transport, the construction of the storage and distribution systems and the securing of the use environments must be respected. The laboratories must be equipped with adequate ventilation, air circulation and alarm systems for gas leaks, and have a first-aid cabinet with relevant medication and therapeutic materials. Training of personnel in artificial breathing techniques is recommended.

In laboratories where their use is necessary, it is prohibited to smoke, use open flames and install explosion-proof electrical equipment and installations. Extinguishers and firefighting outlets must be easily reachble with their signs in sight. For individual safety, gas masks with specific filters and protective clothing must be available.

Gases harmful to the human organism by inhalation, ingestion or skin contact are corrosive and toxic. The asphyxiating effect of toxic gases is due to causticity or chemical reaction. Coming into contact with exposed parts of the body (skin, eyes, respiratory tract), caustic gases (eg $Cl_2$, HF, HCl, $SO_2$, $NH_3$) damage the mucous membranes of the respiratory tract, causing a reflex stop inhalation resulting in suffocation. Reactive gases, such as CO and HCN, are instead combined with the hemoglobin in blood, stopping the $O_2$ intercellular exchange, with the effect of asphyxiating cells. In this case, artificial respiration is essential. Inert gases, such as $N_2$, noble gases, $CO_2$, have no toxic action on the body, but can create asphyxiated conditions if excess in the atmosphere, because they lower the concentration of $O_2$ in the air to below 16%, breathing limit for humans.

### 2.3.2.1. Gas Cylinders

There are two types of compressed gas cylinders. Non-rechargeable cylinders are designed for single use and must never be refilled or reused. The rechargeable cylinders are made of steel or aluminum and are designed for repeated refilling and use. Most reusable cylinders have an open interior with 1/4 inch steel walls and a reinforced neck and bottom. Since the commonly used gases are highly pressurized (between 1,500 psig and 2,500 psig), the cylinders must be kept in good condition and protected from accidental damage at all times. When not in use, the valve is covered with a steel cap, which protects it from shocks. The cylinders shall be used only for the gas for which they are intended for and shall be periodically revised, consisting of an internal control, up tp the capacity of to the volume for filling with water (litres of $H_2O$), in the control of tare and in a test of pressure.

Depending on its critical temperature, the gas at room temperature is in the cylinders compressed in the gaseous state (permanent gases) or in the liquid state in equilibrium with the saturated steam phase. The cylinders must be loaded for about 2/3 of the test pressure (approx. 300 atm for compressed gases). For liquefied gases, the test and the load limit depend on their critical temperature. The internal pressure of compressed gases decreases proportionally to consumption, which can be then controlled by a manometer, while in liquefaction, the internal pressure determined by their saturated vapor pressure at the operating temperature, it remains constant until the total vaporization of the liquid: the residual content can therefore only be known for weighing. The top and the inclined part up to the neck of each gas cylinder are painted the colour assigned to the gas it contains. In the case of a cylinder containing more than one gas, the colours must be applied in such a way as to allow to see each colour seen from above.

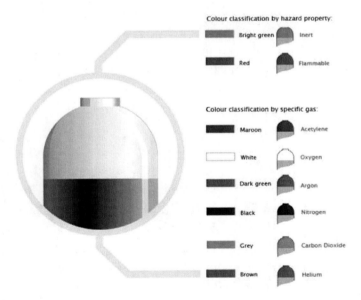

Figure 4. Colour lassification of gas cylinders.

> **Standards for safe use of gas cylinders**
> - If not adequately instructed refrain from the use of gas;
> - Be certain that cylinders are labelled correctly: the type of gas contained must be identifiable by the colour of the ogive;
> - Each cylinder must be fixed securely to the rack;
> - Make sure that the valves are closed when the cylinders are not in use;
> - Carry out the transport of the cylinder with the appropriate trolleys;
> - Always keep cylinders in well ventilated areas;
> - Do not force or attempt to repair the safety valves or the pressure reducers;
> - Do not lubricate valves or reducers with any petroleum product;
> - The empty cylinders must be marked, closed and stored in the deposit with the same precautions as the full ones;
> - The empty cylinders must be subjected to periodic tests: the test date must be printed on the ogive together with the other legal indications.

Figure 5. How to use gas cylinders in safety.

Sometimes the body of the cylinder can be painted with the colour of the major gas and the shoulder the colour of the lesser one. An international chromatic code has been adopted to help identify the gases (Figure 4).

The distribution lines take the gas from the cylinders in the depots located outside the building and organize the distribution in the areas of use: their design depends on the danger and the number of cylinders, the distance of the deposits and the arrangement of the laboratory.

A generic type of line with maximum safety starts from the cylinder, fixed in its storage compartment, with the gearbox connected to the bleed and retention valve equipped with valves from a flexible or metallic spiraltube and continues to the laboratory with a tube equipped with valves of interception and line pressure.

In the laboratory the line branches off to the use ramps, with the control and measurement devices on wall or bench panels, and includes a manual interception valve (with or without manometer) or a reducer with bleed for each branch, and a valve flame retardant calibrated for combustible gases.

The hosepipes must be of of compatible material and impervious to gases; i.e., copper for inert gases (nitrogen, noble gases), oxidizing agents (air, oxygen), hydrogen and hydrocarbons, with the exception of acetylene, and stainless steel, valid for acetylene and for ultrapure gases.

## 2.3.2.2. Pure Gas Generators for Laboratory Use

Gases of extreme purity grade (99.999% or higher) are obtained with laboratory gas production systems, therefore no reduction valves or safety distribution lines are necessary, which are indispensable for compressed gases.

Other advantages are the continuous supply without cylinder replacement, reduced maintenance and low operating costs. The disadvantage of low production capacity has been overcome by the development of new gas generators able to meet the production needs of most laboratory equipment. There is still a restriction on the number of generable gases, limited to hydrogen, nitrogen, air and ozone, while for others (e.g., $H_2S$, $CO_2$, $NH_3$) it is still necessary to resort to traditional Kipp-type appliances.

- *Hydrogen:* $H_2$ has multiple applications in GC, both for the operation of FID, NPD, FPD, ECD, and transport gas detectors. Its use as a transport gas in capillary GC has been limitated by its flammability, compared to the more expensive He, despite his best physical properties, which allow to accelerate separation without loss of efficiency. With the introduction of $H_2$ generators this type of application is possible with technical, economical and safety advantages.

- *Air and nitrogen:* the dry air is composed of about 78% nitrogen, 21% oxygen, 0.9% argon, 0.03% carbon dioxide, traces of methane and other gaseous hydrocarbons and rare gases (neo, helium, crypto, xenon). In the different environmental atmospheres there are also traces of gases and vapors of inorganic and volatile organic substances of various kinds, dispersions of dust in fumes and liquids in the mists. The extraction from the atmosphere of laboratory gases with different degrees of purity requires the removal of unwanted components with suitable separation techniques. For the many uses of $N_2$ in the laboratory, both as transport gas in GC and make-up for ECD, in LC/MS, in thermal analysis, in AAS, both for the creation of high purity inert atmospheres, generators provide for the separation of the gas from the other components of the atmospheric air. The pure air generator is based on the same principle as that for nitrogen, which consists in compressing the ambient air and filtering through materials suitable to retaining particles, water vapours and polluting hydrocarbons, without altering its main gas composition. Purity is of great importance for the use of air as a combustion agent in hydrogen flame detectors (FIDs) in GC.

- *Ozone* is obtained by means of a bench or wall model, which generates ozone continuously or intermittently through a high-voltage and high-frequency energy source. $O_2$ can be used as a gas supply, as well as compressed air or dry air.

## 2.3.3. Laboratory Fume Hoods[*]

Attention should be paid to choosing the position of fume hoods in the laboratory. A wrong choice, placement and use can, in fact, cancel or at least reduce the benefits of its use. The choice of a hood must take into account its intended use, in particular if it will be installed in a chemical or microbiological laboratory [2]. In order to guarantee the safety of the operators in the laboratory over time, it must also:

1. be constructed of suitable materials (metal and non-plastic or wood);
2. possess transparent glass and non-plastic parts;
3. be equipped with suitable filters;
4. have a filter with a thickness of at least 100 mm;
5. have a monitoring system for the depletion and substitution of the filter.

The correct positioning of the extractor hood in the laboratory must take into account in particular the circulation of air flows in the environment and the practical needs of the operator. Hoods located near doors, open windows

---

[*] Chapter 6 on collective protection devices deals with this issue in more detail.

or air conditioning can not provide adequate performance. Hoods should not be in correspondence with the people's passage because the air movements caused can adversely affect the air flow of the hood. If the hoods must be equipped with external expulsion systems, the length of the ducts and the elbows must be kept to a minimum.

According to good laboratory practices, each hood must be subjected to a periodic check and be equipped with the safety register, in which all the determinant elements to guarantee a correct functioning are noted over time:

- Name of the manufacturer, serial number, date of installation and testing;
- Type of filters installed;
- Name of the Company responsible for the hood, name of the maintenance manager, and all data relating to maintenance work.

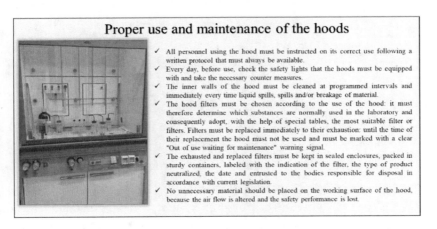

Figure 6. General rules for the use of the hood in safety and its maintenance.

Finally, it is important to check:

1. The efficacy and efficiency of air ejection from hoods (cleaning, periodic filter maintenance, flow control);
2. that the chimneys are positioned one meter beyond the lid;

3. that there are no rules of comparison between incompatible gaseous discharges;
4. that discharges are conveyed directly to the outside or in suitable wells outside the building or in specially designed cavities with suitable fire resistance.

## 2.3.4. Ventilation and Air Change

Attention must be paid to the air conditioning system and the number of air changes in assessing the risk, in order to avoid or at least reduce occupational exposure to hazardous substances by inhalation.

Safety in a laboratory depends on various factors, both environmental and related to specific processes. In general, the nature and quantities of the manipulated substances must be considered, as well as the room cubes, the efficiency of the air conditioning system, the ventilation of the rooms, the correct functioning of the air filtration system.

For a good ventilation system, structural features must be taken:

- air from outside in "unpolluted" areas;
- the division of area into blocks, so that any release of toxic gases (which can not be intercepted with automatic fire dampers) can not reach different environments than the training one;
- tair ventilation speed must be maintained at 0.1-0.2 m/s;
- the system must have fire dampers and an instantaneous fan stop device in an emergency event.

The air conditioning system has the function of maintaining the hygienic-sanitary conditions appropriate for a safe working activity. Several organizations at the international level indicate that air changes in room 4-12/hour is normally adequate general ventilation if the local exhaust systems such as hoods are used as the primary method of control.

The American Association of Industrial Hygienists (ACGIH) underlines the fact that it is misleading to define standard values of air changes/hour. As U.S. OSHA (Occupational Safety and Health Administration) supports, as a ventilation system designer may not know all possible laboratory operations, the chemicals to be used and their potential for the release of fumes and other toxic agents, an air exchange rate (air changes per hour) cannot be specified to meet all conditions. In addition, air changes per hour is not the appropriate concept for the design of contaminant control systems. Contaminants must be checked at source. The control of hazardous chemicals by dilution alone, in the absence of adequate laboratory chemical hoods, is rarely effective in protection of laboratory users. Since unloading from that type of system must be discharged externally or treated intensively before being used as return air, these systems are usually not cheap to control exposure to hazardous materials compared to the use of local exhaust hoods. In each analysis laboratory a plan should be developed to monitor the efficiency of the conditioning system.

## 2.4. Laboratory Equipment

### 2.4.1. Glassware

Glassware are one of the most commonly used equipment in sciencific laboratories. The glassware, before and during washing, must be considered contaminated and dangerous; if broken or cracked, it should not be reused, but must be thrown into a specially marked container to indicate its contents.

This is so that housekeeping and other personnel will exercise due caution during handling. Heat or rapid temperature variations, including cooling, should only be applied to borosilicate, e.g., Pyrex, glassware designed for such applications [3]. Laboratory glassware must be vented when heating to avoid overpressurization and possible explosion. Keep stoppers loose when autoclaving or during procedures that require heating. Conventional vacuum glass dryers represent a potential risk of explosion/implosion and it is therefore advisable to replace them with plastic resin dryers that do not produce fragments in the event of an accident. In the lower part, any desiccant agent (with the exception of concentrated acids) is applied; inside there is an unbreakable plate resistant to temperature variations, where the samples will be positioned. The manufacturer's instructions must be complied with for cleaning. Proper handling of glassware reduce the risk of injury and accident [4].

### General Precautions on the safe handling of lab glassware

- Before using any piece of glassware, always take time to examine it carefully and ensure that it is in good condition. Do not use any glassware that is scratched, chipped, cracked or etched. Defects like these can seriously weaken the mechanical strength of the glass and cause it to break in use.
- Dispose of broken or defective glassware safely. Use a purpose-designed disposal bin that is puncture resistant and clearly labelled. Pyrex glassware (or any other borosilicate glass) should under no circumstances be disposed of in a domestic glass recycling stream, as its high melting point makes it incompatible with other glass (soda-lime glass) for recycling. The correct method of disposal is to include it in the general waste in accordance with the relevant guidelines, provided that the glass is free from any harmful chemical contamination.
- Never use excessive force to fit rubber bungs into the neck of a piece of glassware. Always ensure that you select the correct size of bung.
- Carrying or lifting large glass flasks, beakers or bottles, etc. by the neck or rim can be very dangerous. Always provide support from the base and sides.
- When stirring solutions in glass vessels, avoid using stirring rods with sharp ends which can scratch the glassware causing it to become weakened.
- Always heat glassware gently and gradually to avoid sudden temperature changes which may cause the glass to break due to thermal shock. Similarly, allow hot glassware to cool gradually and in a location away from cold draughts.
- If you are using a hotplate, ensure that the top plate is larger than the base of the vessel to be heated. If the base of the vessel overhangs the hotplate top, hotspots can occur causing the base of the vessel to break. Also, never put cold glassware onto a pre-heated hotplate. Always warm up the glassware from ambient temperature.
- If you are using a Bunsen burner, employ a soft flame and use a wire gauze with a ceramic center to diffuse the flame. Never apply direct localized heat to a piece of glassware.
- Pyrex borosilicate glass is microwave safe. However, as with any microwave vessel, ensure that it holds microwave absorbing material, before placing it in the oven.
- When autoclaving containers, always loosen off the caps. Autoclaving glassware with a tightly screwed cap can result in pressure differences which will cause the container to break.
- Always wear thick protective gloves and safety spectacles. Never use force.
- Carefully rock the cone in the socket to achieve separation.
- If the joint is dry, try to provide lubrication. Hold the joint upright and add penetrating oil to the top of the cone. Wait until the penetrating oil is well dispersed into the joint before trying to separate.

Figure 7. Rules for the handling of laboratory glassware.

### How to use the centrifuge safely

- Make sure that the centrifugal bowls and tubes are dry.
- Make sure the spindle is clean.
- Use groups of tubes, buckets and other matched equipment and inspect them for cracks or defects before use, avoiding overfilling.
- Make sure that the tubes or containers are adequately balanced in the rotor.
- Always use cups of safety centrifuges to hold spills and prevent aerosols.
- Make sure the rotor is correctly positioned on the driveshaft.
- If you are adequately trained, check the O-rings on the rotor.
- Apply vacuum grease in accordance with the manufacturer's guidelines.
- Do not exceed the maximum travel speed of the rotor.
- Close the centrifuge cover during operation.
- Make sure that the centrifuge is operating normally before leaving the area and that the rotor has reached a complete stop before opening the lid.

Figure 8. Procedures for the safe use of the centrifuge.

## 2.4.2. Centrifuges

The serious hazards associated with centrifugation include mechanical failure and the creation of aerosols. Only trained users must operate centrifuges. Must be installed properly according to the manufacturer's recommendations. It is important that the load is balanced each time the centrifuge is used and that the lid is closed while the rotor is moving. The disconnect switch must work properly to turn off the equipment when the hop is open, and the manufacturer's instructions must be followed for safe operating speeds. The Laboratory Safety Manager must ensure that all personnel using a centrifuge are able to minimize the potential risks. For the disinfection of centrifuges 0.5% sodium hypochlorite is used.

## 2.4.3. Autoclave

The autoclave is a common tool in laboratories, particularly in microbiology laboratories, and is a source of danger for operators if it is not used correctly (the use procedure must be fixed near the machine). The autoclave must only be used by authorized personnel and after wearing appropriate PPE (face mask, heat protection gloves). Physical hazards involve heat, vapour and pressure, biological hazards involve potential exposure to viable human pathogens:

1. contact with the overheated surface;
2. contact with superheated steam to open the lid without having respected the right waiting times;
3. contact with the steam coming out of the vent valve;
4. failure of the safety valve: the instrument is led to too high pressure;
5. contact with biocontaminated material for incomplete sterilization of products (the turning of the sterilization indicator must always be checked before removing the contaminants).

34 *Maria Pia Gatto*

Sterilization will only occur when the conditions of time, temperature, pressure and humidity are met. Incorrect selection of the time or exhaust cycle can damage the autoclave, cause liquid to boil over, or bottles to break. The correct use of autoclave will minimize the possibility of serious injury.

---

### Procedures for Using Steam Autoclaves

✓ NEVER AUTOCLAVE FLAMMABLE, REACTIVE, CORROSIVE, TOXIC or RADIOACTIVE MATERIALS, e.g., bleach.
✓ Materials that melt (plastic lab wear) at ≥ 121°C will block chamber exhaust drain if not placed in a shallow autoclave pan able to withstand that temperature.
✓ Use caution when increasing autoclave temperature to 135°C because plastics (including some plastic pans) melt at this temperature, causing difficult clean-up and damage to temperature sensors.
✓ Always wear safety glasses, goggles, or face shield, lab coat or apron, and heat-protective non-asbestos gloves when opening door or removing item(s) from autoclave.
✓ Do not mix loads that require different exposure times and exhaust.
✓ Open door only after chamber pressure returns to zero. Leave door open for several minutes to allow pressure to equalize and for materials to cool.
✓ Open door slowly. Beware of rush of steam or water.

---

Figure 9. General rules for a safe use of autoclave.

## 2.4.4. Refrigerators

The refrigerators in the laboratory must be thawed and cleaned periodically. According to an established programme, the internal temperature must be monitored and reported in a special document, to be stored in the workplace.

Reagents, samples and solutions that must be stored at a controlled temperature (range 0; +5°C) must be cataloged and labeled and tightly closed. Never store food and beverages in refrigerators where reagents and specimens are stored.

## 2.4.5. Mittens

The incineration of samples requires the attainment of temperatures above 500°C and, consequently, the adoption of safety measures to reduce the risk of burns.

## The Analytical Laboratory

- Only the staff can access the muffle furnace; the area around the flask must be bordered by yellow lines indicating the area of risk;
- Insert a sign with the words "Warning High Temperatures" in good condition when the flask is in operation;
- Replace old-fashioned mitts with new ones with insulated walls and a "non-tipping" but "folding" door so that the risk of burns is not sustained when the crucibles are removed from the incineration chamber;
- Equip the operators with laboratory gloves for anti-scalding;
- Write an operating procedure which clearly indicates the operations to be followed in the use of the flask, the PPE to be worn, the interventions in the event of an accident;
- Place a first aid kit near the muffle in case of burning.

### Table 10. Gravity displacement steam autoclave in the biomedical laboratory*

| Items | Biological Waste (Gravity Cycle) | Liquids (Liquid Cycle) | Dry Items (Gravity Cycle) | Glassware (Gravity Cycle) |
|---|---|---|---|---|
| Preparation | Open the bag >2", Place in tray, Place indicator if needed | Loosen caps or use a vented closure, Fill containers no more than 75% capacity | Fabrics Wrap; Instruments: Clean, dry, lay in pan | Dirty: Place in middle of the pan; Clean: wash, rinse, wrap |
| Placement in Autoclave | In the center | Upright in pan | Fabrics: Separated, on edge; Instruments: Flat | Dirty: In detergent and pan; Clean: On side or inverted |
| Temperature | 121°C | 121°C | 121°C | 121°C |
| Treatment Time in Minutes | 60-120 min. depending on load size and packing density | 22 min for volumes | 100mL 30-60 min | 30-60 min |
| Exhaust Cycle | Slow exhaust | Slow exhaust | Fast exhaust and dry | Dirty: Slow exhaust; Clean: Fast/dry |
| Notes | Avoid puncturing bags. Overbag and dispose of properly. | Hot bottles may explode. Let cool before moving. | Check reference for proper packaging methods | Glassware with cracks or deep scratches may crack |

*Reference: "Using the Gravity Displacement Steam Autoclave in the Biomedical Laboratory" DHHS/PHS/HIH/DS. Source: https://www.drs.illinois.edu/site-documents/ Autoclave%20Poster-Final%20(11x17).pdf.

## 2.4.6. Working with Burners, Flames and Hotplates

One of the risks that must be taken into due consideration in the laboratory is the flame produced by gas. Open flames in the laboratory are discouraged due to a potential fire hazard, especially when working with flammable materials.

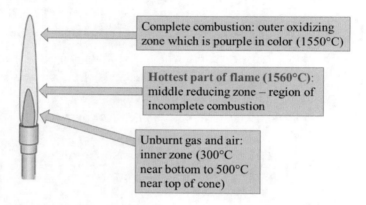

Figure 10. Bunsen burner flame temperatures.

If you use a Bunsen burner:

- At the time of purchase, make sure that the Bunsen burner meet the required requirements by issuing from the supplier the documentation attesting to compliance with current regulations;
- During installation, use reinforced connecting pipes to ensure tightness and ensure that it is solidly affixed to the Bunsen burner and the gas supply source. It is dangerous to leave a Bunsen burner on unattended;
- Place the instrument not in correspondence with sources of light which could make the flame not well visible with the risk of burn to the operator. Do not store matchboxes in the immediate vicinity;
- Use only self-igniting bunsen. The Bunsen lamp must be scrupulously subjected to routine maintenance. The possible interventions can be daily (cleaning operations, checking the

tightness of the fastening of the gas supply pipe clamps), every six months (checking the integrity of the connection pipe), annually (revision by the customer assistance service);

- Place hotplates away from any flammable materials, such as alcohol, acetone, and toluene, and bottles of reagents. Hotplates can spark, unseen by your eyes, and ignite flammable liquids or vapors. Hotplates retain heat for a while after they have been used and turned off. Once used, hotplates must be treated as if they were hot, even after disconnecting them from the power supply. Before moving hotplate and putting it away, give enough time to cool it;
- If you need to leave the lab, please inform your collegues and your instructor that the hotplate, left on the lab bench, is still hot.

## 2.4.7. Safe Handling of Products

In laboratories there is not always staff specifically responsible for carry out storage, distribution and therefore daily transport of the necessary quantities of product in the various sectors of the laboratory, as well as repositioning in the depots of the same products; but it is often the same operators that perform the above tasks.

Inside the laboratories, the products must be distributed with trolleys with sides that prevent the fall of the products and have a fallarrest tank in case of breakage of the containers. The transport of individual glass containers is not advisable. A proper procedure may involve transporting the glass bottle inside a plastic container with handles.

### 2.4.7.1. Handling of Liquid Nitrogen

Nitrogen has a boiling point of -196°C, so under normal pressure conditions, it is in the liquid phase at lower temperatures. In order to avoid cold burns, the operator must pay great attention during the handling, storage, transfer of liquid nitrogen and the objects that have been immersed in it.

| Safely use of liquid nitrogen ||
|---|---|
| To protect **skin** you must:<br>✓ operate in well-ventilated areas;<br>✓ always wear a gown, suitable gloves and a face mask;<br>✓ do not wear trousers with the cuff;<br>✓ never fill the containers with liquid nitrogen up to the top;<br>✓ never hermetically close the containers for liquid nitrogen and allow the vapors to escape;<br>✓ take great care not to cause liquid nitrogen to escape during transfers;<br>✓ take off the clothes and protective clothing on which liquid nitrogen has been inadvertently poured;<br>✓ carefully read the warnings for the transfer of liquid nitrogen;<br>✓ remove the liquid nitrogen before repairing or intervening on the containers;<br>✓ in case of burns go immediately to the emergency room. | **Atmospheric oxygen dilution**<br>Nitrogen vapor released into closed environment air create a dilution of the oxygen concentration, necessary for breathing. Exposure to this diluted atmosphere can cause asphyxiation. To avoid this, it is necessary not to store or use liquid nitrogen in not sufficiently ventilated environments. <br><br>**Burns**<br>Cryogenic liquids, including condensed nitrogen, must be handled with great care. Skin contact with cryogenic liquids or cold gas can in fact cause burns. During the transfer it is advisable to always protect the eyes and the skin. For the protection of the arms wear insulated long gloves; the trousers must be worn over the shoes in order to protect the operator from accidental spills of liquid. To protect the eyes, wear goggles or safety masks to avoid the risk of splashing cryogenic liquids. |

Figure 11. How to use liquid nitrogen.

### 2.4.8. UV Germicidal Lamps

Germicidal lamps are used in a variety of applications where disinfection is the main concern, including the purification of air and water, the protection of food and beverages and the sterilization of sensitive instruments such as medical instruments. Germicidal light destroys the ability of bacteria, viruses, and other pathogens to multiply by deactivating their reproductive capacities.

The germicidal lamps emit radiation in the UV-C portion of the ultraviolet (UV) spectrum, which includes wavelengths between 100 and 280 nanometers (nm). The UV radiation (UVR) used in most germicidal bulbs is harmful to both skin and eyes, and germicidal bulbs should not be used in any fixture or application that has not designed specifically to prevent exposure to humans or animals. UVR does not feel immediately; in fact, the user may not realize the danger until after the exposure caused damage.

Symptoms typically occur from 4 to 24 hours after exposure.

The Analytical Laboratory 39

Exposure of inadequately shielded ultraviolet operators may result:

- *Acute effects* appear within hours of exposure, while chronic effects are long-lasting and cumulative and may not appear for years. An acute effect of UVR is redness of the skin called erythema (similar to sunburn);
- *Chronic effects* include accelerated skin aging and skin cancer.

## 2.5. THE TWENTY-FIVE MOST IMPORTANT SAFETY RULES IN THE LABORATORY

1. Access to non-professionals is prohibited;
2. Eating, drinking and smoking is prohibited in the laboratory;
3. All exit routes and areas around safety equipments (showers, eyewash, fire extinguishers ...) must be clear from obstructions;
4. The laboratory floor must be kept free of obstacles (electric cables, boxes, ...), cleaned of residues (glass shards, ...) and dry;
5. Read the safety data sheet carefully before using any product;
6. Do not stay to work alone in the laboratory: accidents always occur without notice and can be fatal without immediate help;
7. Verify that all containers are correctly labeled;
8. Never leave current reactions or unattended operating equipment. Nighttime preparations should be avoided. However, if a reaction is to continue during the night, the control must be entrusted to special safety devices which switch off the power supply following significant variations in temperature, flow or level beyond certain limits;
9. Never place containers, bottles or appliances near the edge of the workbench;
10. Communicate with your colleagues to notify the tests in progress if hazardous substances are managed;

11. Check the efficiency and effectiveness of the cooling devices during operations;
12. The broken glassware shall be stored in the appropriate safety containers;
13. Do not attempt to solve the problem yourself, but inform the head of the laboratory in case of malfunctions of machines, equipment and systems;
14. Do not pipet with the mouth;
15. The use of gloves outside laboratories is strictly forbidden; do not touch the door handles and other laboratory objects with gloves that have been manipulated with samples or reagents;
16. Do not hold scissors, steel spatulas, glass tubes or blunt instruments in your pockets;
17. Avoid the use of contact lenses: if you can not do without them, you must at all times wear safety glasses;
18. Refrigerators with flammable substances stored must have an explosion-proof electrical system, the prohibition to introduce flammable materials must be highlighted on the door of the refrigerator with an ordinary electrical system;
19. All operations with hazardous substances must be carried out under the hood;
20. All chemicals must be eliminated by hazardous waste procedures;
21. All accidents (even those resolved without damage or injury) must be reported to the laboratory manager and recorded in the laboratory notebook, highlighting causes and emergency interventions;
22. Do not leave unidentifiable material in the work areas;
23. Open shoes and high-heeled shoes should not be used in the laboratory as weel as long hair should be kept collected; dangling jewels (earrings, bracelets, etc.) may be accident risk factors;
24. Learn well all the rules of first aid: in case of emergency, every passing minute can have serious consequences;
25. Use appropriate personal protective equipment (PPE) in the laboratory for each level of risk (clothing, disposable gloves,

goggles, suitable protective masks, shoes). The PPE must be correctly used and kept in a good state of maintenance.

# REFERENCES

[1] Cornell University EH&S. *Cornell University EH&S Laboratory Safety Manual and Chemical Hygiene Plan Engineering Controls* (2015, March). Retrieved from https://sp.ehs.cornell.edu/lab-research-safety/laboratory-safety-manual/Pages/ch2.aspx

[2] University of Bristol Health and Safety Office (July 2012). *University of Bristol Health and Safety Office Fume Cupboard Guidance* (2015, March). Retrieved from http://www.bristol.ac.uk/safety/media/ gn/fume-cupboards-gn.pdf

[3] University of Vermont, Safety in Laboratories, *Identifying the Hazards: Safe Handling of Glassware*. (2015, March). Retrieved from http://www.uvm.edu/safety/lab/safe-handling-of-glassware

[4] Zakharyuta, Anastasia and Şen, Canhan and Avaz, Merve Senem and Akkaş, Tuğçe and Pürçüklü, Sibel and Baytekin Birkan, Tuğba and Gönül, Turgay and Yerdelen, Bilge and Cebeci, Fevzi Çakmak and İnce, Adnan) Sabanci University, Istanbul. (2016) *Laboratory Safety Handbook*. ISBN 978-605-9178-58-7.

*Chapter 3*

# CHEMICAL RISK

## 3.1. THE CHEMICALS THAT WE CAN FIND IN AN ANALYTICAL LAB

Hazardous chemicals present physical and/or health threats to workers in clinical, industrial, and academic laboratories [1]. Hazardous laboratory chemicals include carcinogens [2], toxins that may affect the liver, kidney or nervous system, irritants, corrosives, and sensitizers, as well as agents acting on the blood system or damaging lungs, skin, eyes, or mucous membranes. OSHA [3] rules limit all industry exposures to approximately 400 substances [4].

Laboratories where employees may be exposed to hazardous chemicals must develop and implement a written chemical hygiene plan and designate a chemical hygiene officer to ensure compliance with OSHA standard 1910.1450 [5]. This standard generally applies to all chemical laboratory activities and specifies that employee exposures to hazardous chemicals be at or below the PELs and TLVs.

Chemical substances and mixtures are identifiable as:

- not dangerous;
- not dangerous but used in conditions that could constitute a hazard (e.g., high temperature water, compressed gases); in these cases the risk is not chemical, as it is linked to the physical characteristics of the agent (pressure, temperature, etc.);
- hazardous but not classified as such by the legislation (intermediate reaction substancies, fumes which may develop in the progress of activities, etc.);
- classified and labeled as hazardous by existing regulations, with appropriate symbols, risk phrases and precautionary instructions specified on the label. The dangers are conventionally divided into three groups: physical hazards, health hazards and environmental hazards. Within the same class of danger further subdivisions into categories describe the risk graduality and therefore the probability of adverse effects.

## 3.2. UNDERSTANDING CHEMICAL HAZARDS

There are three main "exposure pathways," or ways in which a chemical can enter the body:

- *Respiration (inhalation):* breathe in chemical gases, mists, or powders that are in the air;
- *Contact with skin or eye:* getting chemicals on the skin, or in the eyes. They can damage the skin, or be absorbed through the skin into the bloodstream;
- *Swallowing (ingestion):* this can happen when the chemicals have spilled or settled on food, beverages, cigarettes, beards or hands. Once chemicals have entered your body, some can move into your bloodstream and reach the "target" internal organs, such as the lungs, the liver, the kidneys or the nervous system.

The effects of a toxic chemical on your body may be either acute or chronic.

Acute (short-term) effects occour immediately after exposure to the chemical. They may be minor, like nose or throat irritation, or they could be serious, like eye damage or passing out from chemical vapors. What all these effects have in common is that they happen right away.

Chronic (long-term) effects may take years to present. They are usually caused by regular exposure to a harmful substance over a long period of time. These effects are usually permanent. Some chemicals cause both acute and chronic effects. For example, breathing solvent vapors could cause dizziness immediately (an acute effect). But breathing the same vapors all the time for many years could possibly cause liver damage (a chronic effect).

## 3.3. DEFINITION OF OCCUPATIONAL EXPOSURE LIMITS

An occupational exposure limit, an important tool in risk assessment and the management of activities involving the manipulation of dangerous substances, is an upper limit to the acceptable concentration of a hazardous substance in the workplace air for a particular material or class of materials [6 ].

## Definitions:

- A *Permissible Exposure Limit (PEL)* is the maximum amount or concentration of a chemical that a worker may be exposed to under OSHA regulations;
- A *Threshold Limit Value (TLV)* of a chemical substance is the level to which a worker can be exposed day after day for a working lifetime without adverse effects (American Conference of Governmental Industrial Hygienists). Three types of TLVs for chemical substances are defined:

- *TLV-TWA, Time-Weighted Average:* average exposure on the basis of a 8h/day, 40h/week work schedule.
- *TLV-STEL, Short-Term Exposure Limit:* spot exposure for a duration of 15 minutes, that cannot be repeated more than 4 times per day with at least 60 minutes between exposure periods.
- *TLV-C Ceiling:* absolute exposure limit that should not be exceeded at any time.

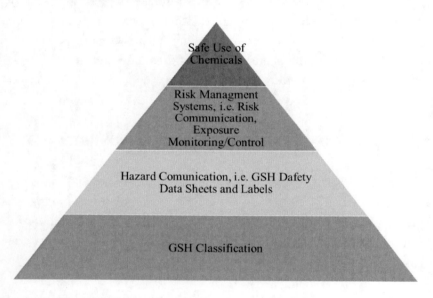

Figure 12. The GHS* pyramid of safety chemicals management. *GSH is an international system that the United Nations created for the unified classification and labeling of chemicals, officially adopted in the U.S. on March 26, 2012, by the Occupational Safety and Health Administration (OSHA) called HazCom 2012.

Figure 12 shows the foundation of programs to ensure the safe use of chemicals, according to the GHS (Globally Harmonized System) of Classification and Labelling of Chemicals [7].

The first two steps in any program to ensure the safe use of chemicals are to identify the intrinsic hazard(s) (i.e., classification) and thus to communicate such information. The design of the GHS communication elements reflect the different needs of various target audiences, such as workers and consumers.

# Table 11. Classification of hazardous substances and mixtures

| Physical Hazards |
|---|
| Explosives An explosive substance (or mixture) is a solid or liquid which is in itself capable by chemical reaction of producing gas at such a temperature and pressure and at such a speed as to cause damage to the surroundings. Pyrotechnic substances are included even when they do not evolve gases. |
| Flammable Gases Flammable gas means a gas having a flammable range in air at 20°C and a standard pressure of 101.3 kPa. |
| Flammable Aerosols Aerosols are any gas compressed, liquefied or dissolved under pressure within a non-refillable container made of metal, glass or plastic, with or without a liquid, paste or powder. |
| Oxidizing Gases Oxidizing gas means any gas which may, generally by providing oxygen, cause or contribute to the combustion of other material more than air does. |
| Gases Under Pressure Gases under pressure are gases that are contained in a receptacle at a pressure not less than 280 Pa at 20°C or as a refrigerated liquid. This endpoint covers four types of gases or gaseous mixtures to address the effects of sudden release of pressure or freezing which may lead to serious damage to people, property, or the environment independent of other hazards the gases may pose. |
| Flammable Liquids Flammable liquid means a liquid having a flash point of not more than 93°C. Substances and mixtures of this hazard class are assigned to one of four hazard categories on the basis of the flash point and boiling point. |
| Flammable Solids Flammable solids are solids that are readily combustible, or may cause or contribute to fire through friction. |
| Self-Reactive Substances Self-reactive substances are thermally unstable liquids or solids liable to undergo a strongly exothermic thermal decomposition even without participation of oxygen (air). |
| Pyrophoric Liquids A pyrophoric liquid is a liquid which, even in small quantities, is liable to ignite within five minutes after coming into contact with air. |
| Pyrophoric Solids A pyrophoric solid is a solid which, even in small quantities, is liable to ignite within five minutes after coming into contact with air. |
| Self-Heating Substances A self-heating substance is a solid or liquid, other than a pyrophoric substance, which, by reaction with air and without energy supply, is liable to self-heat. This endpoint differs from a pyrophoric substance in that it will ignite only when in large amounts (kilograms) and after long periods of time (hours or days). |
| Substances that, in contact with water, emit flammable gases Solids or liquids which, by interaction with water, are liable to become spontaneously flammable or to give off flammable gases in dangerous quantities. |
| Oxidizing Liquids An oxidizing liquid is a liquid which, while in itself not necessarily combustible, may, generally by yielding oxygen, cause or contribute to the combustion of other material. |
| Oxidizing Solids An oxidizing solid is a solid which, while in itself not necessarily combustible, may, generally by yielding oxygen, cause or contribute to the combustion of other material. |
| Organic Peroxides An organic peroxide is an organic liquid or solid which contains the bivalent -0-0- structure and may be considered a derivative of hydrogen peroxide, where one or both of the hydrogen atoms have been replaced by organic radicals. |
| Corrosive to Metals A substance or a mixture that by chemical action will materially damage, or even destroy, metals is termed 'corrosive to metal'. |

## Table 11. (Continued)

| Health Hazards |
| --- |
| Acute Toxicity Substances are assigned to one of the five toxicity categories on the basis of LD50 (oral, dermal) or LC50 (inhalation). The LC50 values are based on 4-hour tests in animals. The GHS* provides guidance on converting 1-hour inhalation test results to a 4- hour equivalent. |
| Skin Corrosion/Irritation Skin corrosion means the production of irreversible damage to the skin following the application of a test substance for up to 4 hours. Substances and mixtures in this hazard class are assigned to a single harmonized corrosion category. |
| Serous Eye Damage/Eye Irritation Skin irritation means the production of reversible damage to the skin following the application of a test substance for up to 4 hours. Substances and mixtures in this hazard class are assigned to a single irritant category. |
| Respiratory or Skin Sensitization *Respiratory sensitizer* means a substance that induces hypersensitivity of the airways following inhalation of the substance. *Skin sensitizer* means a substance that will induce an allergic response following skin contact. The definition for "skin sensitizer" is equivalent to "contact sensitizer". |
| Germ Cell Mutagenicity An agent giving rise to an increased occurrence of mutations in populations of cells and/or organisms. |
| Carcinogenicity A chemical substance or a mixture of chemical substances which induce cancer or increase its incidence. |
| Reproductive Toxicology Reproductive toxicity includeds adverse effects on sexual function and fertility in adult males and females, as well as developmental toxicity in offspring. |
| Target Organ Systemic Toxicity (TOST) Some existing systems distinguish between single and repeat exposure for these effects and some do not. All significant health effects, not otherwise specifically included in the GHS, that can impair function, both reversible and irreversible, immediate and/or delayed are included in the nonlethal target organ/systemic toxicity class (TOST). Narcotic effects and respiratory tract irritation are considered to be target organ systemic effects following a single exposure. |
| Aspiration Toxicity Severe acute effects such as chemical pneumonia, varying degrees of pulmonary injury or death following aspiration. Aspiration is the entry of a liquid or solid directly through the oral or nasal cavity, or indirectly from vomiting, into the trachea and lower respiratory system. |
| * GHS= the United Nations Globally Harmonized System of Classification and Labeling of Chemicals. |
| **Enviromental Hazards** |
| Acute Acute aquatic toxicity means the intrinsic property of a material to cause injury to an aquatic organism in a short-term exposure. Chronic aquatic toxicity means the potential or actual properties of a material to cause adverse effects to aquatic organisms during exposures that are determined in relation to the lifecycle of the organism. |

To proceed further up the pyramid, some existing national programs also include risk management systems as part of an comprehensive programme on sound chemical management. The general goal of these systems is to minimize exposure, resulting in reduced risk [8]. The systems vary in focus and include activities such as establishing exposure limits, recommending exposure monitoring methods and creating engineering controls. However, the target audiences of such systems are generally limited to workplace settings. With or without formal risk management systems, the GHS is designed to promote the safe use of chemicals.

## 3.4. THE SAFETY DATA SHEET (SDS)

The use of chemicals and mixtures in the laboratory leads the operator in contact with reduced volumes of substances, which are often used in the mixture, thus producing exposures to more agents, but generally at low doses. In the laboratory, chemical agents may be numerous, but their identification is now made much faster and more exhaustive by the obligation imposed on manufacturers and suppliers to affix special labels on containers and to attach to products (Material) Safety Data Sheets (SDS), which must report all the required information for the use of the product in maximum safety conditions [9]. The SDS provides comprehensive information for use in the management of chemical substancies in the workplace by supplying the necessary information on chemical, physical and toxicological properties, together with suggestions for storage, transportation and disposal of chemicals. SDSs are the best source of information available and should be consulted as a first step in assessing the risk associated with the experiment or working with a new product.

Employers and workers use the SDS as sources of hazard information and to obtain advice on safety precautions. The SDS is related to the product and, usually, is not able to provide specific information for any given workplace where the product can be used. However, information from the SDS allows the employer to develop an active programme of protection measures for workers, including training, which is specific to the individual workplace and to take into account all measures that may be necessary to protect the environment [10]. The information conteined in a SDS also provides a source of information for other target audiences such as those involved in the transport of dangerous goods, rescuers, poison control centers, those involved in the professional use of pesticides and consumers.

**Table 12. GHS pictograms and hazard classes**

| Explosives Self Reactives Organic Peroxides | Flammables Self Reactives Pyrophorics Self-Heating Emits Flammable Gas Organic Peroxides | Oxidizers | Acute toxicity (severe) | Corrosives |
|---|---|---|---|---|
| Gases Under Pressure | Irritant Dermal Sensitizer Acute toxicity (harmful) Narcotic Effects Respiratory Tract Irritation | Carcinogen Respiratory Sensitizer Reproductive Toxicity Target Organ Toxicity Mutagenicity Aspiration Toxicity | Environmental Toxicity | |

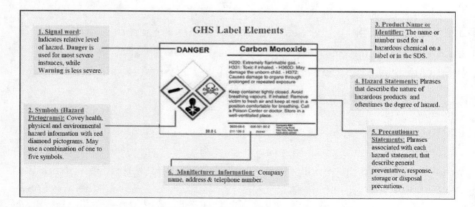

Figure 13. The six elements required by HCS on labels of hazardous chemicals.

## Table 13. List of H codes and EU supplementary hazard statements list

| |
|---|
| H200: Unstable explosive |
| H201: Explosive; mass explosion hazard |
| H202: Explosive; severe projection hazard |
| H203: Explosive; fire, blast or projection hazard |
| H204: Fire or projection hazard |
| H205: May mass explode in fire |
| H206: Fire, blast or projection hazard; increased risk of explosion if desensitizing agent is reduced |
| H207: Fire or projection hazard; increased risk of explosion if desensitizing agent is reduced |
| H208: Fire hazard; increased risk of explosion if desensitizing agent is reduced |
| H220: Extremely flammable gas |
| H221: Flammable gas |
| H222: Extremely flammable aerosol |
| H223: Flammable aerosol |
| H224: Extremely flammable liquid and vapour |
| H225: Highly flammable liquid and vapour |
| H226: Flammable liquid and vapour |
| H227: Combustible liquid |
| H228: Flammable solid |
| H229: Pressurized container: may burst if heated |
| H230: May react explosively even in the absence of air |
| H231: May react explosively even in the absence of air at elevated pressure and/or temperature |
| H232: May ignite spontaneously if exposed to air |
| H240: Heating may cause an explosion |

## Table 13. (Continued)

| |
|---|
| H241: Heating may cause a fire or explosion |
| H242: Heating may cause a fire |
| H250: Catches fire spontaneously if exposed to air |
| H251: Self-heating; may catch fire |
| H252: Self-heating in large quantities; may catch fire |
| H260: In contact with water releases flammable gases which may ignite spontaneously |
| H261: In contact with water releases flammable gas |
| H270: May cause or intensify fire; oxidizer |
| H271: May cause fire or explosion; strong oxidizer |
| H272: May intensify fire; oxidizer |
| H280: Contains gas under pressure; may explode if heated |
| H281: Contains refrigerated gas; may cause cryogenic burns or injury |
| H290: May be corrosive to metals |
| H300: Fatal if swallowed |
| H301: Toxic if swallowed |
| H302: Harmful if swallowed |
| H303: May be harmful if swallowed |
| H304: May be fatal if swallowed and enters airways |
| H305: May be harmful if swallowed and enters airways |
| H310: Fatal in contact with skin |
| H311: Toxic in contact with skin |
| H312: Harmful in contact with skin |
| H313: May be harmful in contact with skin |

| |
|---|
| H314: Causes severe skin burns and eye damage |
| H315: Causes skin irritation |
| H316: Causes mild skin irritation |
| H317: May cause an allergic skin reaction |
| H318: Causes serious eye damage |
| H319: Causes serious eye irritation |
| H320: Causes eye irritation |
| H330: Fatal if inhaled |
| H331: Toxic if inhaled |
| H332: Harmful if inhaled |
| H333: May be harmful if inhaled |
| H334: May cause allergy or asthma symptoms or breathing difficulties if inhaled |
| H335: May cause respiratory irritation |
| H336: May cause drowsiness or dizziness |
| H340: May cause genetic defects |
| H341: Suspected of causing genetic defects |
| H350: May cause cancer |
| H351: Suspected of causing cancer |
| H360: May damage fertility or the unborn child |
| H361: Suspected of damaging fertility or the unborn child |
| H361d: Suspected of damaging the unborn child |
| H362: May cause harm to breast-fed children |
| H370: Causes damage to organs |
| H371: May cause damage to organs |
| H372: Causes damage to organs through prolonged or repeated exposure |
| H373: May cause damage to organs through prolonged or repeated exposure |

**Table 13. (Continued)**

| |
|---|
| H400: Very toxic to aquatic life |
| H401: Toxic to aquatic life |
| H402: Harmful to aquatic life |
| H410: Very toxic to aquatic life with long-lasting effects |
| H411: Toxic to aquatic life with long-lasting effects |
| H412: Harmful to aquatic life with long-lasting effects |
| H413: May cause long-lasting harmful effects to aquatic life |
| H420: Harms public health and the environment by destroying ozone in the upper atmosphere |
| EUH001: Explosive when dry. |
| EUH006: Explosive with or without contact with air. |
| EUH014: Reacts violently with water. |
| EUH018: In use, may form flammable/explosive vapour-air mixture. |
| EUH019: May form explosive peroxides. |
| EUH044: Risk of explosion if heated under confinement. |
| EUH029: Contact with water liberates toxic gas. |
| EUH031: Contact with acids liberates toxic gas. |
| EUH032: Contact with acids liberates very toxic gas. |
| EUH066: Repeated exposure may cause skin dryness or cracking. |
| EUH070: Toxic by eye contact. |
| EUH071: Corrosive to the respiratory tract. |
| EUH059: Hazardous to the ozone layer. |
| EUH201: Contains lead. Should not be used on surfaces liable to be chewed or sucked by children. |
| EUH201A: Warning! Contains lead. |

| |
|---|
| EUH202: Cyanoacrylate. Danger. Bonds skin and eyes in seconds. Keep out of the reach of children. |
| EUH203: Contains chromium (VI). May produce an allergic reaction. |
| EUH204: Contains isocyanates. May produce an allergic reaction. |
| EUH205: Contains epoxy constituents. May produce an allergic reaction. |
| EUH206: Warning! Do not use together with other products. May release dangerous gases (chlorine) |
| EUH207: Warning! Contains cadmium. Dangerous fumes are formed during use. See information supplied by the manufacturer. Comply with the safety instructions. |
| EUH208: Contains (name of sensitising substance). May produce an allergic reaction. |
| EUH209: Can become highly flammable in use. |
| EUH209A: Can become flammable in use. |
| EUH210: Safety data sheet available on request. |
| EUH401: To avoid risks to human health and the environment, comply with the instructions for use. |

**Table 14. ADR Pictograms. Class 6.2 (infectious substances), class 7 (radioactive material) and class 9 (miscellaneous dangerous substances and articles) pictograms are included in the UN Model Regulations but have not been incorporated into the GHS because of the nature of the hazards**

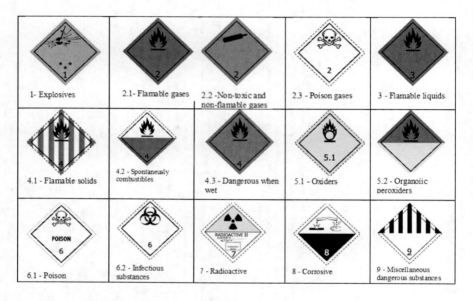

The SDS should contain 16 headings:

1. *Identification of the substance or mixture and of the supplier:* GHS product identifier. Other means of identification. Recommended use of the chemical and restrictions on use. Supplier's details (including name, address, phone number, etc.). Emergency phone number;
2. *Hazards identification:* GHS classification of the substance/mixture and any national or regional information. GHS label elements, including precautionary statements. (Hazard symbols may be provided as a graphical reproduction of the symbols in black and white or the name of the symbol, e.g., flame, skull and crossbones.) Other hazards, which do not result in classification (e.g., dust explosion hazard) or are not covered by the GHS.

The GHS hazard statement means a standard phrase assigned to a hazard class and category to describe the nature and severity of a chemical hazard. Each hazard statement is designated a code, starting with the letter H and followed by 3 digits.

- H2xx: Physical hazards;
- H3xx: Health hazards;
- H4xx: Environmental hazards.

The H code is used for reference purpose only. It is the actual phrase which should appear on labels and safety data sheets.

Labels, as defined in the Standard of Hazard Communication (HCS), are an appropriate group of written, printed or graphic informational elements concerning a hazardous chemical that are affixed to, printed or attached to the immediate container of a hazardous chemical or to the outer packaging. HCS requires chemical manufacturers, importers or distributors to ensure that any container of hazardous chemicals coming out of the workplace is labeled, tagged or marked with the following information: product identifier; signal word; hazard statement(s); precautionary statement(s); and pictogram(s); and name, address and telephone number of the chemical manufacturer, importer or other responsible party.

In most cases, the precautionary statements are independent. However, OSHA allows flexibility to apply precautionary declarations to the label, such as combining declarations, using an order of precedence, or deleting an inappropriate statement. Precautionary instructions can be combined on the label to save space and improve readability. For example, "Keep away from heat, spark and open flames," "Store in a well-ventilated place," and "Keep cool" may be combined to read: "Keep away from heat, sparks and open flames and store in a cool, well-ventilated place." Where a chemical is classified for a number of hazards and the precautionary statements are similar, the most stringent statements must be included on the label. In this case, the manufacturer, importer or chemical distributor may impose an order of precedence in which the response phrases require rapid action to ensure the health and safety of the person exposed.

Employers are responsible for maintaining labels on containers, including, but not limited to, tanks, totes and drums. This means that labels must be kept on chemicals in a manner that continues to be readable and that the relevant information (such as the hazards and directions for use) is not laid (i.e., vanish, washed) or removed in any way.

3. *Composition/information on ingredients*: Substance - Chemical identity; Common name, synonyms, etc; CAS number, EC number, etc; Impurities and stabilizing additives which are themselves classified and which contribute to the classification of the substance. Mixture - The chemical identity and concentration or concentration ranges of all ingredients which are hazardous within the meaning of the GHS and are present above their cutoff levels.

4. *First aid measures*: Description of necessary measures, subdivided according to the different routes of exposure, i.e., inhalation, skin and eye contact, and ingestion. Most important symptoms/effects, acute and delayed. Indication of immediate medical attention and special treatment needed, if necessary.

5. *Firefighting measures*: Suitable (and unsuitable) extinguishing media. Specific hazards arising from the chemical (e.g., nature of any hazardous combustion products). Special protective equipment and precautions for firefighters.

6. *Accidental release measures*: Personal precautions, protective equipment and emergency procedures. Environmental precautions. Methods and materials for containment and cleaning up.

7. *Handling and storage*: Precautions for safe handling. Conditions for safe storage, including any incompatibilities.

8. *Exposure controls/personal protection*: Control parameters, e.g., occupational exposure limit values or biological limit values. Appropriate engineering controls. Individual protection measures, such as personal protective equipment.

9. *Physical and chemical properties*: Appearance (physical state, color, etc.); odor; odor threshold; pH; melting point/freezing point; initial boiling point and boiling range; flash point; evaporation rate;

flammability (solid, gas). Upper/lower flammability or explosive limits; vapor pressure; vapor density; relative density; solubility(ies); partition coefficient: n-octanol/water; autoignition temperature; decomposition temperature.

10. *Stability and reactivity*: Chemical stability; Possibility of hazardous reactions; Conditions to avoid (e.g., static discharge, shock or vibration); Incompatible materials; Hazardous decomposition products. A wide variety of chemicals react dangerously when mixed with certain other materials. Some of the more widely-used incompatible chemicals are given at the end of this paragraph (Table 15).

11. *Toxicological information*: Concise but complete and comprehensible description of the various toxicological (health) effects and the available data used to identify those effects, including: information on the likely routes of exposure (inhalation, ingestion, skin and eye contact). Symptoms related to physical, chemical and toxicological characteristics. Delayed and immediate effects and also chronic effects from short- and long-term exposure. Numerical measures of toxicity (such as acute toxicity estimates).

12. *Ecological information*: Ecotoxicity (aquatic and terrestrial, where available); Persistence and degradability; Bioaccumulative potential; Mobility in soil; Other adverse effects.

13. *Disposal considerations*: Description of waste residues and information on their safe handling and methods of disposal, including the disposal of any contaminated packaging.

14. *Transport information*: UN Number; UN Proper shipping name; Transport Hazard class(es); Packing group, if applicable; Marine pollutant (Yes/No); Special precautions which a user needs to be aware of or needs to comply with in connection with transport or conveyance either within or outside their premises.

15. *Regulatory information*: Safety, health and environmental regulations specific for the product in question.

16. *Other information including information on preparation and revision of the SDS.*

60 *Maria Pia Gatto*

## Table 15. Incompatible lab chemicals (not exhaustive list)

| Chemical Substance | Incompatible Agents |
|---|---|
| acetic acid | chromic acid, ethylene glycol, nitric acid, hydroxyl compounds, perchloric acid, peroxides, permanganates |
| acetone | concentrated sulphuric and nitric acid mixtures |
| acetylene | chlorine, bromine, copper, fluorine, silver, mercury |
| alkali and alkaline earth metals | water, chlorinated hydrocarbons, carbon dioxide, halogens, alcohols, aldehydes, ketones, acids |
| aluminium (powdered) | chlorinated hydrocarbons, halogens, carbon dioxide, organic acids. |
| anhydrous ammonia | mercury, chlorine, calcium hypochlorite, iodine, bromine, hydrofluoric acid |
| ammonium nitrate | acids, metal powders, flammable liquids, chlorates, nitrites, sulphur, finely divided organic combustible materials |
| aniline | nitric acid, hydrogen peroxide |
| arsenic compounds | reducing agents |
| azides | acids |
| bromine | ammonia, acetylene, butadiene, hydrocarbons, hydrogen, sodium, finely-divided metals, turpentine, other hydrocarbons |
| calcium carbide | water, ethanol |
| calcium oxide | water |
| carbon, activated | calcium hypochlorite, oxidizing agents |
| chlorates | ammonium salts, acids, metal powders, sulphur, finely divided organic or combustible materials |
| chromic acid | acetic acid, naphthalene, camphor, glycerin, turpentine, alcohols, flammable liquids in general |
| chlorine | see bromine |
| chlorine dioxide | ammonia, methane, phosphine, hydrogen sulfide |
| copper | acetylene, hydrogen peroxide |
| cumene hydroperoxide | acids, organic or inorganic |
| cyanides | acids |
| flammable liquids | ammonium nitrate, chromic acid, hydrogen peroxide, nitric acid, sodium peroxide, halogens |
| hydrocarbons | fluorine, chlorine, bromine, chromic acid, sodium peroxide |
| hydrocyanic acid | nitric acid, alkali |
| hydrofluoric acid | aqueous or anhydrous ammonia |
| hydrogen peroxide | copper, chromium, iron, most metals or their salts, alcohols, acetone, organic materials, aniline, nitromethane, flammable liquids, oxidizing gases |
| hydrogen sulphide | fuming nitric acid, oxidizing gases |
| hypochlorites | acids, activated carbon |
| iodine | acetylene, ammonia (aqueous or anhydrous), hydrogen |

| Chemical Substance | Incompatible Agents |
| --- | --- |
| mercury | acetylene, fulminic acid, ammonia |
| mercuric oxide | sulphur |
| nitrates | sulphuric acid |
| nitric acid (conc.) | acetic acid, aniline, chromic acid, hydrocyanic acid, hydrogen sulphide, flammable liquids, flammable gases |
| oxalic acid | silver, mercury |
| perchloric acid | acetic anhydride, bismuth and its alloys, ethanol, paper, wood |
| peroxides (organic) | acids, avoid friction or shock |
| phosphorus (white) potassium | air, alkalies, reducing agents, oxygen carbon tetrachloride, carbon dioxide, water, alcohols, acids |
| potassium chlorate | acids |
| potassium perchlorate | acids |
| potassium permanganate | glycerin, ethylene glycol, benzaldehyde, sulphuric acid |
| selenides | reducing agents |
| silver | acetylene, oxalic acid, tartaric acid, ammonium compounds, fulminic acid |
| sodium | carbon tetrachloride, carbon dioxide, water |
| sodium nitrate | ammonium salts |
| sodium nitrite | ammonium salts |
| sodium peroxide | ethanol, methanol, glacial acetic acid, acetic anhydride, benzaldehyde, carbon disulfide, glycerin, ethylene glycol, ethyl acetate, methyl acetate, furfural |
| sulphides | acids |
| sulphuric acid | potassium chlorate, potassium perchlorate, potassium permanganate (or compounds with similar light metals, such as sodium, lithium, etc.) |
| tellurides | reducing agents |
| zinc powder | sulphur |

# 3.5. SAFE STORAGE OF CHEMICALS IN LABORATORIES

A range of storage facilities is suitable for chemicals in the laboratory environment; many of these are specially designed for safe storage of different types of hazardous substances. It is important to understand what substances can be safely stored in which storage container.

*Flammable solvent cabinets* are made of metal or wood with a minimum fire resistance of a half hour (some are to one and a half hour standard).

They should contain a spillage tray made of suitable material that is compatible with solvents.

*Ventilated cabinets* are fitted with forced ventilation. They may be free-standing with their own extract system, or may be situated beneath a fume cupboard and attached to its duct. They are designed to safely store chemicals that give off noxious fumes and smells. These fumes are sucked away by the forced ventilation.

*Refrigerators & freezers* may be used for the preservation of certain hazardous substances, however, if the substances are flammable, the unit must not contain any internal light source or thermostat which may provide a source of ignition for any flammable vapour. Refrigerators and laboratory freezers that meet these requirements are available from the main laboratory supply companies, domestic appliances should be avoided.

Storage depots for materials must meet certain essential requirements that will be summarized in the list below [11]:

- the deposit must be above the ground and physically separated from the laboratory;
- floors must be waterproof and non-sparking if flammable materials are stored;
- the storage of flammable liquids must take place in a room separate from those where the other reagents are stored or, for small quantities, in a fire resistant safety cabinet and with the possibility of containment in case of breakage;
- cables must be present for the antistatic bonding of metallic containers to be used during the transfer of flammable liquids;
- the deposit must be sufficiently large so that you can buy multiple small capacity containers;
- facility must be present to circumscribe, collect, neutralize any spills;
- containers and materials must be labeled properly;
- highly reactive materials must be stored separately;

# Chemical Risk

- highly toxic materials such as $H_2S$, $CN^-$, $Cl_2$, $CO$ must be labelled, stored and manipulated in such a way that no one can be exposed to risk;

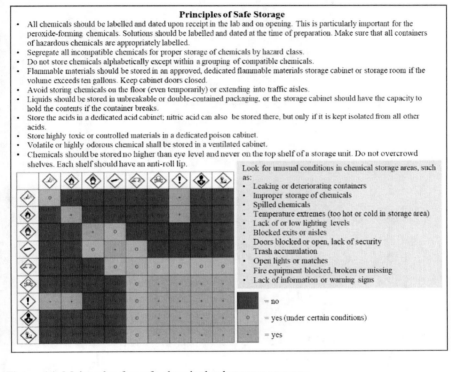

Figure 14. Main rules for safe chemical substances storage.

- flammable materials must be purchased in small-capacity metallic containers;
- procedures should be provided for the collection and storage of flammable liquids and hazardous chemical wastes;
- compressed, flammable and toxic gas cylinders are deposited outside the building, securely and protected from the elements;
- flammable gas cylinders must be kept separate from those of the oxiders;
- the cylinders must always be moved with special trolleys fitted with fasteners and must be anchored in the place of use;

- the pipes and fittings must be recognizable and readily identifiable by the fluid being transported;
- explosive products, such as organic peroxides, are stored in areas far from flammable liquids.

## 3.6. WASTE MANAGEMENT

The characterization, management storage and disposal of laboratory wastes (i.e., chemical waste including hazardous and non-hazardous solid waste, radioactive or mixed waste, biohazardous and medical waste, and universal waste) requires strict compliance with regulatory obligations. Definitions:

- Biological Waste – Biological materials of animal, human, plant or microbial origin;
- Biohazardous waste – This can include infectious material, contaminated agar plates, live cultures, human cells and blood, and disposables that have been in contact with the above;
- Broken Glass bin – These are the yellow bins located within certain labs. The contents of these bins are placed into the domestic waste stream, so should not contain any hazardous material.
- Contaminated Materials Lab Bin – Bins located in labs for glass, filter paper, tissues, gloves which are not free of hazardous material;
- Sharps – Objects or devices that have acute, rigid corners, edges, points or protuberances capable of cutting or penetrating the skin e.g., hypodermic needles, glass, scalpel blades and lancets. All sharps are hazardous because of the potential to cause cuts and punctures;
- Winchester – Brown glass container, most often 2.5 L in volume, used to store solvents, ammonia solution and concentrated acids.

# Chemical Risk 65

Waste chemicals can take various forms including solvents, aqueous solutions, dry powders, and unwanted old chemicals. The SDS for each chemical must be checked to ensure compatibility of materials for mixed chemical residue containers.

**Laboratory Sharps Disposal**

"Sharps waste" means any device having acute rigid corners, edges, or protuberances capable of cutting or piercing, including, but not limited to, all of the following: hypodermic needles, syringes, razor blades and scalpel blades. Glass items contaminated with biohazards, such as pipettes, microscope slides and capillary tubes are also considered a sharps waste*.

**Bin color: yellow (general) or purple (cytotoxic) approved sharps container**
**Final disposal method: dependent upon primary contamination**

Sharps are objects or devices that have acute, rigid corners, edges, points or protuberances capable of cutting or penetrating the skin e.g. hypodermic needles, broken glass, scalpel blades, lancets, syringes with needles, razor-blades.
Place any sharps in approved sharps container. Sharps containers should be located adjacent to the work area where sharps are used for easy access. Sharps may also be contaminated with toxic, infectious or radioactive materials. These sharps should be placed into separate sharps containers which are then labelled appropriately according to the type of primary contamination (chemically contaminated, biohazardous, radioactive, and cytotoxic). When the sharps residue container is filled to the "fill line", seal container, affix Hazardous Waste label (if required), and place in appropriate wheelie bin.
Do not overfill container - only fill to "fill line" marked on container.

*AS 4031-1992: Non-reusable containers for the collection of sharp medical items used in health care areas.
AS 4261-1994: Reusable containers for the collection of sharp items used in human and animal medical applications.
AS 4478-1997: Guide to the reprocessing of reusable containers for the collection of sharp items used in human and animal clinical/medical applications.

Figure 15. Laboratory sharps rolules for final disposal. Sharps may also be contaminated with toxic, infectious or radioactive materials, which substantially increase the risk potential.

When possible, mixing of chemicals should be avoided to prevent unexpected reactions. The waste container should be compatible with the residue material placed inside. If the wastes are liquid, the residue containers are approved solid plastic sealable containers. No hazardous chemical substances should be disposed down drains. Generally chemical waste must be segregated according to its properties, such as:

- *Class 1 (Explosive) and Class 4 (Spontaneously Combustible)* wastes cannot be disposed of by the routes mentioned in Table 15.
- *Halogenated solvent* wastes are to be collected in waste containers, labelled as halogenated solvents. Halogenated wastes must be kept separate to other organic solvents as, for example, mixtures of acetone and chloroform can explode;

- *Cyanide wastes* must be placed in an appropriate waste bottle and the solution kept alkaline at all times;
- *Dry chemicals* should be placed in a drum (not more than 5L), labelled "Waste Chemicals for Disposal". Strong oxidising and reducing agents (chlorates, bromates, peroxides, nitrates, iodides, metal dusts, hypochlorites, etc.) should not be placed in this drum. See your supervisor for instructions on the disposal of these reactive dry chemicals. They should never be placed with organic chemicals;
- *Highly reactive substances* such as amines, phosphorus compounds, acetic anhydride, acetyl- chloride should never be placed in general disposal containers;
- *Dilute, non-toxic chemicals* may be washed, after waste supervisor approval, into the sewerage system. Large quantities of any chemical should be returned to the store for recovery or disposal;
- *Not down the drain*! Volatile solvents and smelly substances can enter the drains via rotary evaporators, vacuum filtration or from carelessness. Have you ever had a smell arising from the drain in your lab? Please think about what is going down your drain - especially drains in fume-hoods as you maybe unaware of the smell that is escaping. Everyone should make sure that traps are used and refilled regularly to prevent hazardous materials entering the drains and waterways.

## REFERENCES

[1] American Chemical Society, *Safety in the Academic Chemistry Laboratories*, 4th edition, 1985.

[2] IARC *Monographs on the Evaluation of the carcinogenic Risk of Chemicals to Man*, World Health Organization Publications Center, 49 Sheridan Avenue, Albany, New York 12210 (latest editions).

## Chemical Risk

[3] *NIOSH/OSHA Pocket Guide to Chemical hazards*, NIOSH Pub. No. 85-11, U.s. Government Printing Office, Washington, DC, 1985 (or latest edition).

[4] Hill, R.H. Jr., and Finster, D.C. (2010) *Laboratory Safety for Chemistry Students*, John Wiley & Sons, Inc., New York.

[5] OSHA 1910.1450 - *Occupational exposure to hazardous chemicals in laboratories*.

[6] Stricoff, S.R., and Walters, D.B. (1995) *Handbook of Laboratory Health and Safety*, 2nd edition, John Wiley & Sons, Inc., New York.

[7] *Globally Harmonized System of Classification and Labelling of Chemicals* (Second revised ed.), (2007). New York and Geneva: United Nations, ISBN 978-92-1-116957-7, ST/SG/AC.10/30/Rev.2 ("GHS Rev.2").

[8] Young, J.A., (1987) Ed., *Improving Safety in the Chemical Laboratory*, John Wiley & Sons, Inc. New York.

[9] Kenkel, J., (2000) *Lab Safety Training Chemistry: An Industry-Based Laboratory Manual*, CRC Press, Inc.,

[10] National Research Council, *Prudent Practices for Handling Hazardous Chemicals in Laboratories*, National Academy Press, Washington, DC, 2010.

[11] Furr, K.A. (2000) *CRC Handbook of Laboratory Safety*, 5th edition, CRC Press, Inc., New York.

*Chapter 4*

# BIOLOGICAL RISK

## 4.1. BIOLOGICAL HAZARDS

Workers may be exposed to biological agents if they are in contact with:

- natural or organic materials such as soil, plant materials (hay, straw, cotton, etc.)
- substances of animal origin (wool, hair, etc.)
- food
- organic dust (e.g., flour, paper dust, animal dander)
- waste, dirty water
- blood and other body fluids.

In general, we can define:

- *biological agents:* microorganisms, including those that have been genetically modified, cell cultures and human endoparasites, which may be able to cause any infection, allergy or toxicity;
- *micro-organism:* a microbiological entity, cellular or non-cellular, capable of replication or of transferring genetic material;
- *cell culture:* the in-vitro growth of cells derived from multicellular organisms [1].

## 4.2. Biological Risk Management

One of the most helpful tools available to carry out a microbiological risk assessment is the list of risk groups for microbiological agents [2]. However, a simple reference to the risk grouping for a given agent is insufficient in the execution of a risk assessment [3].

Other factors that should be considered, if appropriate, include:

- Pathogenicity of the agent and infectious dose;
- Potential outcome of exposure;
- Natural route of infection;
- Other routes of infection, resulting from laboratory manipulations (parenteral, airborne, ingestion);
- Stability of the agent in the environment;
- Concentration of the agent and volume of concentrated material to be manipulated;
- Presence of a suitable host (human or animal);
- Information available from animal studies and reports of laboratory-acquired infections or clinical reports;
- Laboratory activity planned (sonication, aerosolization, centrifugation, etc.);
- Any genetic manipulation of the organism that may extend the host range of the agent or alter the agent's sensitivity to known, effective treatment regimens;
- Local availability of prophylaxis or effective therapeutic interventions.

On the basis of the information ascertained during the risk assessment, it is possible to assign a level of biosafety to the planned work, the appropriate personal protective equipment selected and standard operating procedures (SOPs) which incorporate other safety interventions developed to ensure the safest possible conduct of the work.

Any strain known more dangerous than the parent (wild-type) strain should be considered for handling at a higher containment level. Certain attenuated strains or strains that have been demonstrated to have irreversibly lost known virulence factors may qualify for a reduction of the containment level compared to the Risk Group assigned to the parent strain.

Steps in a biological risk assessment:

- Identify the hazards associated with the biological agent and assess the risk. The risk of the agent can include the ability to infect and cause disease, the severity of disease, the route of exposure, the medical surveillance and the treatment availability;
- Review the laboratory procedures and the hazards associated with these procedures. Procedural hazards may include concentration of the agent, volume, equipment used, aerosol generating procedures, use of sharps and hazards associated with working with animals;
- Make a determination of the appropriate biosafety level. Additional precautions may be required on the basis of risk;
- Evaluate staff regarding their technical skills and the safety equipment;
- Review the risk assessment with experienced persons and with the Institutional Committee for Biosafety.

## 4.3. CLASSIFICATION AND LABELING

The World Health Organization (WHO) recommended a classification of risk groups of laboratory use agents, describing four general risk groups based on these main characteristics and on the route of transmission of natural disease.

These four groups address the risk for both the laboratory worker and the community. The *NIH Guidelines* have established a comparable classification and human etiological agents assigned in four risk groups on the basis of the hazard (Table 16).

## Table 16. Biological risks classification groups

| Risk Group Classification | NIH Guidelines for Research involving Recombinant DNA Molecules 2002[1] | World Health Organization Laboratory Biosafety Manual 3rd Edition 2004[2] |
|---|---|---|
| Risk Group 1 | Agents not associated with disease in healthy adult humans. | (No or low individual and community risk) A microorganism unlikely to cause human or animal disease. |
| Risk Group 2 | Agents associated with human disease that is rarely serious and for which preventive or therapeutic interventions are often available. | (Moderate individual risk; low community risk) A pathogen that can cause human or animal disease but is unlikely to be a serious hazard to laboratory workers, the community, livestock or the environment. Laboratory exposures may cause serious infection, but effective treatment and preventive measures are available and the risk of spread of infection is limited. |
| Risk Group 3 | Agents associated with serious or lethal human disease for which preventive or therapeutic interventions may be available (high individual risk but low community risk). | (High individual risk; low community risk) A pathogen that usually causes serious human or animal disease but does not ordinarily spread from one infected individual to another. Effective treatment and preventive measures are available. |
| Risk Group 4 | Agents likely to cause serious or lethal human disease for which preventive or therapeutic interventions are not usually available (high individual risk and high community risk). | (High individual and community risk)A pathogen that usually causes serious human or animal disease and can be readily transmitted from one individual to another, directly or indirectly. Effective treatment and preventive measures are not usually available.[3] |

[1] The National Institutes of Health (US), Office of Biotechnology Activities. NIH guidelines for research involving recombinant DNA molecules. Bethesda; 2002, April.
[2] World Health Organization. Laboratory biosafety manual. 3rd ed. Geneva; 2004.
[3] American Public Health Association. Control of communicable diseases manual. 18th ed. DL Heymann, editor. Washington, DC; 2005.

Risk groups correlate with but do not equate to biosafety levels (BSL). A risk assessment will determine the degree of correlation between an agent's risk group classification and biosafety level [4]. In this regard, the main characteristics of the four BLS are listed below:

## Biosafety Level 1 (BSL-1)

As the lowest of the four, biosafety level 1 applies to laboratory settings in which personnel work with low-risk microbes that pose little to no threat of infection in healthy adults. An example of a microbe that is typically worked with at a BSL-1 is a nonpathogenic strain of *E. coli*.

This laboratory setting typically consists of research taking place on benches without the use of special contaminant equipment. A BSL-1 lab, which is not required to be isolated from surrounding facilities, houses activities that require only standard microbial practices, such as:

- Mechanical pipetting only (no mouth pipetting allowed);
- Safe sharps handling;
- Avoidance of splashes or aerosols;
- Daily decontamination of all work surfaces when work is complete;
- Hand washing;
- Prohibition of food, drink and smoking materials in lab setting;
- Personal protective equipment, such as eye protection, gloves and a lab coat or gown;
- Biohazard signs.

BSL-1 labs also requires immediate decontamination after spills. Infection materials are also decontaminated prior to disposal, generally through the use of an autoclave.

## Biosafety Level 2 (BSL-2)

This level of biosafety includes laboratories working with agents associated with human diseases (i.e., pathogenic or infections organisms) that pose a moderate health hazard. Examples of agents typically worked with in a BSL-2 include equine encephalitis viruses and HIV, as well as *Staphylococcus aureus (staph infections)*.

BSL-2 laboratories maintain the same standard microbial practices as BSL-1 labs, but also includes enhanced measures due to the potential risk of the aforementioned microbes. Personnel working in BSL-2 labs should take care even more to prevent injuries such as cuts and other breaches of the skin, as well as ingestion and mucous membrane exposures.

In addition to BSL 1 expectation, the following practices are required in a BSL 2 lab setting:

- Appropriate personal protective equipment (PPE) must be worn, including lab coats and gloves. Eye protection and face shields can also be worn, as needed;
- All procedures that can cause infection from aerosols or splashes are performed within a biological safety cabinet (BSC);
- An autoclave or an alternative method of decontamination is available for proper disposals;
- The laboratory has self-closing, lockable doors;
- A sink and eyewash station should be readily available;
- Biohazard warning signs.

Access to a BSL-2 lab is far more restrictive than a BSL-1 lab. Outside personnel, or those with an increased risk of contamination, are often restricted from entering when work is being conducted.

## Biosafety Level 3 (BSL-3)

Again building upon the two prior biosafety levels, a BSL-3 laboratory typically includes work on microbes that are either indigenous or exotic, and can cause serious or potentially lethal disease through inhalation. Examples of microbes worked with in a BSL-3 includes; yellow fever, West Nile virus, and the bacteria that causes tuberculosis.

The microbes are so serious that the work is often strictly controlled and registered with the appropriate government agencies. Laboratory personnel are also under medical surveillance and could receive immunizations for microbes they work with. Common requirements in a BSL-3 laboratory include:

- Standard personal protective equipment must be worn, and respirators might be required;
- Solid-front wraparound gowns, scrub suits or coveralls are often required;
- All work with microbes must be performed within an appropriate BSC;
- Access hands-free sink and eyewash are available near the exit;
- Sustained directional airflow to draw air into the laboratory from clean areas towards potentially contaminated areas (Exhaust air cannot be re-circulated);
- A self closing set of locking doors with access away from general building corridors.

Access to a BSL-3 laboratory is restricted and controlled at all times.

## Biosafety Level 4 (BSL-4)

BSL-4 labs are rare. However some do exist in a small number of places in the US and around the world. As the highest level of biological safety, a BSL-4 lab consists of working with highly dangerous and exotic microbes. Infections caused by these types of microbes are frequently fatal and come without treatment or vaccines. Two examples of such microbes include Ebola and Marburg viruses.

In addition to BSL-3 considerations, BSL-4 laboratories have the following containment requirements:

- Personnel are required to change clothing before entering, shower upon exiting;
- Decontamination of all materials before exiting;

76 *Maria Pia Gatto*

- Personnel must wear appropriate personal protective equipment from prior BSL levels, as well as a full body, air-supplied, positive pressure suit;
- A Class III biological safety cabinet.

A BSL-4 laboratory is extremely isolated—often located in a separate building or in an isolated and restricted zone of the building. The laboratory also has a dedicated supply and exhaust air, as well as vacuum lines and decontamination systems.

Knowing the difference in biosafety lab levels and their corresponding protection requirements is essential for anyone working with microbes in a lab environment.

**Table 17. Biosafety levels**

| Biosafety Level | 1 | 2 | 3 | 4 |
|---|---|---|---|---|
| Infectious Agents | Unlikely to cause disease in healthy workers or animals Low individual and community risk | Can cause human or animal disease but unlikely to be a serious hazard Moderate individual risk, limited community risk Effective treatments available | | |
| Examples of infectious agents in this risk level | | E. coli, California encephalitis viruses, many influenza viruses | Anthrax, Q fever, tuberculosis, Hantaviruses, human immuno-deficiency viruses | Ebola viruses, Herpes B virus (Monkey virus), foot and mouth disease |
| Facilities | Standard well-designed experimental animal and laboratory facilities | Level 1 plus: Separate laboratory, room surfaces impervious and readily cleaned, biohazard sign | Level 2 plus: Controlled access double door entry and body shower, air pressure must be negative at all times, no recirculation, HEPA filtration, backup power | Specialized, secure, completely self-contained unit with specialized ventilation, fully monitored; air lock entry and exit |

# Biological Risk

| Biosafety Level | 1 | 2 | 3 | 4 |
|---|---|---|---|---|
| Safety Equipment | Handwashing facilities, laboratory coats | Level 1 plus: autoclave, HEPA filtered class I or II biological safety cabinet, personal protective equipment | Level 2 plus: Autoclave, HEPA filtered class II biological safety cabinet, personal protective equipment to include solid front laboratory clothing, head covers, dedicated footwear, and gloves, appropriate respiratory | Class III biological safety cabinets, positive pressure ventilated suits |
| Procedures | Basic safe laboratory practices | Use of personal protective equipment laboratory coat worn only in the laboratory, gloves, decontamination | Users fully trained, written protocols; showers, wastes disposed of as contaminated, use of biological safety cabinets, personal protective devices | Access only to certified staff, rigorous sterilization/ decontamination procedures |

Adopted from Canadian Council on Animal Care.

## Laboratory Waste Disposal routes

**Cytotoxic Waste**
*Bin color: purple base with purple lid - Final disposal method: incineration.*
Cytotoxic waste is any substance contaminated with any residue or preparations that contain materials that are toxic to cells principally by their action on cell reproduction. All cytotoxic waste (class 6.1) should be placed in an approved purple cytotoxic bag or container. When the residue container is full, place in purple labelled cytotoxic waste wheelie bin kept in secure area.
Although the final disposal method for cytotoxic waste is the same as chemically contaminated waste, it must be treated more securely prior to incineration due its mutagenic potential.

**Biological/Clinical, Biosecurity Waste**
*Bin color: yellow base with yellow lid - Final disposal method: autoclave then landfill.*
Biological/clinical waste must be rendered non-viable before disposal. This generally means autoclaving. Where applicable, any biosecurity waste must be effectively contained and disposed in a manner approved by the Department of Agriculture and Water Resources.
Waste that has been chemically treated must NOT be autoclaved.

**Glass Waste**
*Bin color: green base with brown lid (brown glass e.g. Winchesters) green base with white lid (clear glass) - Final disposal method: recycling.*
Glass, whether broken or unbroken, should not be placed in general waste bins. The bottle cap can be removed and disposed in the general waste bin. Once clean, place glass in waste bin based on glass color. When the laboratory glass bin is ¾ full, the lid should be placed on the bin and the contents transferred to the larger solid waste bins.
Contaminated glass containers or laboratory glass such as beakers, volumetric flasks of other Pyrex items cannot be placed in general recycling bins.

**Radioactive Contaminated Waste**
*Bin color: red base with red lid - Final disposal method: dependent on primary hazard.*
Radioactive waste should be packaged according to its primary hazard e.g. Chemically Contaminated Waste or Biological/Clinical, Biosecurity Waste. It will be kept in the Radioactive Waste Store to "delay and decay" prior to final disposal as non-radioactive waste.

Figure 16. Final disposal rules for laboratory waste matherial.

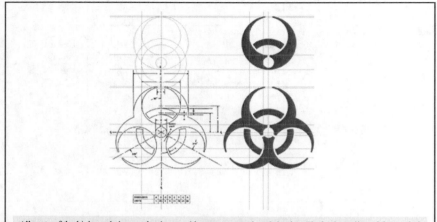

Figure 17. International Biohazard Warning Symbol.

The symbol of biohazard is used in the labeling of biological materials that pose a significant risk to health, biohazard, including viral samples and hypodermic needles used. The international biohazard warning symbol and sign (Figure 16) must be displayed on the doors of the environments where microorganisms of Risk Group 2 or higher risk groups are managed.

## 4.4. BIOLOGICAL RISK PREVENTION

The biohazard assessment aims to identify the different types of hazards related to the manipulation of biological agents in order to eliminate, or reduce to an acceptable level, the risk of contamination of operators, samples, the environment and the community in general. Risk control is carried out through the definition and adoption of appropriate prevention measures such as:

# Biological Risk

- Suitable containment levels (Paragraph 4.4.1.);
- Satisfactory equipment;
- Appropriate rules of behavior in the laboratory (Paragraph 4.4.2.);
- Adequate collective and/or individual protection measures (Chapters 6 & 7).

The professionalism, training, experience and common sense of the operator are also essential for the removal or reduction of the risk of contamination. The training and periodical updating of the personnel and the development of a manual with the indication of appropriate procedures to be adopted during the activities or in the event of an accident are therefore an integral part of the prevention programme.

## 4.4.1. Laboratory Containment Levels

Containment laboratories must be designed and constructed to prevent or control exposure to the biological agent in use of laboratory workers, other people and the environment in general. As we pointed out in the previous chapter, biological agents were categorised into four hazard groups because of their infectivity and the consequences of infection by the Advisory Committee on Dangerous Pathogens (ACDP). The containment levels usually required for work with such agents are determined by their categorisation and these reflect the increasing levels of health risk to those involved in (or who could be affected by) such work. Containment levels define the minimum requirements necessary to provide adequate protection for personnel working with biological agents and to prevent contamination of the surrounding environment.

There are 4 levels of containment depending on the operations carried out within the laboratory.

*Containment level 1:* for laboratory operations involving zero or negligible risks to human health and the environment are performed; minimum containment and protection measures must therefore be applied. Laboratory rooms must be separated from the outside through a door that should remain closed during the activities. The presence of a sink is recommended.

*Containment level 2:* for operations presenting a low risk for human health and the environment. This level of containment requires the presence of a class I or II biological safety hood to protect the worker from possible aerosol formations. The biohazard signal must be exposed on the laboratory door. An autoclave must be present in the laboratory or in the building in order to inactivate the waste before disposal (Figure 18).

*Containment level 3:* for operations that present a moderate risk to human health and the environment. Access to the laboratory is strictly controlled and the presence of a class I or II biological safety cabinet is required. The biohazard signal must be exposed on the laboratory door. An autoclave must be present in the laboratory or in the plane. In the latter case, procedures must be adopted that allow the safe transfer of material to an autoclave outside the laboratory [5]. Access to these laboratories must take place via a filter area (Figure 18). The side of the contamination-free filter zone must be separated from the restricted access part of a changing room or shower facilities and, preferably, from self-locking doors. Inside the laboratories the pressure must be negative. The air emitted from the laboratory must be subjected to ultrafiltration (HEPA). The work area must be sealable in order to allow fumigation.

*Containment level 4:* for high risk to human health and to the environment operations. The laboratory must be separated from other areas of the same building or must be in a separate building. Access to the laboratory is strictly controlled and the presence of a class I, II or III biological safety cabinet is required. A double-entry autoclave must be present. Access to these laboratories must be through a filter area in which a decontamination shower must be present. Inside the laboratories, the pressure must be negative. The air introduced and emitted by the laboratory must be subjected to ultrafiltration (HEPA).

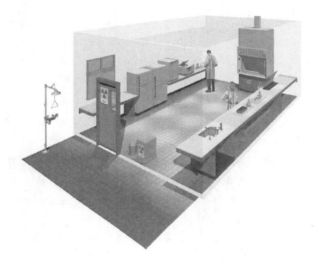

Figure 18. A typical Biosafety Level 2 laboratory (provided by CUH2A, Princeton, NJ, USA).

### 4.4.1.1. Biosafety Levels 1 and 2

Whatever the activities carried out in a laboratory, the following indications (and those related to biosafety levels 3 and 4 in the paragraph 4.4.1.2) taken from the WHO (World Health Organization) indications [6], must be kept in mind:

1. Ample space must be provided for the safe conduct of laboratory work and for cleaning and maintenance.
2. Walls, ceilings and floors should be smooth, easy to clean, impermeable to liquids and resistant to chemicals and disinfectants normally used in the laboratory. Floors should be resistant to slipping.
3. Bench must be waterproof and resistant to disinfectants, acids, alkalis, organic solvents and moderate heat.
4. The lighting should be adequate for all activities. Undesirable reflections and glare should be avoided.
5. Laboratory furniture should be sturdy. Open spaces between and under benches, cabinets and equipment should be accessible for cleaning.

6. Storage space must be sufficient to contain supplies for immediate use and thus avoid confusion on the counter tops and in the aisles. It should also provide a long-term storage space, conveniently located outside the laboratory working spaces.
7. Space and facilities should be provided for the safe handling and storage of solvents, radioactive materials and compressed and liquefied gases.
8. Equipment for storing external clothing and personal items must be provided outside the laboratory working areas.
9. Facilities for eating and drinking and for rest should be provided outside the laboratory working areas.
10. The hand-washing basins, with running water, if possible, should be supplied in each laboratory room, preferably near the exit door.
11. Doors should have vision panels, appropriate fire ratings, and preferably be selfclosing.
12. At Biosafety Level 2, an autoclave or other means of decontamination should be available in appropriate proximity to the laboratory.
13. Safety systems should cover fire, electrical emergencies, emergency shower and eyewash facilities.
14. First-aid areas or rooms suitably equipped and readily accessible should be available.
15. In the planning of new facilities, it should be considered the supply of mechanical ventilation systems that provide an inward flow of air without recirculation. If there is no mechanical ventilation, the windows should be able to be opened and should be equipped with arthropod-proof screens.
16. A reliable supply of good quality water is essential. There should be no crossconnections between sources of laboratory and drinking-water supplies. An antibackflow device should be fitted to protect the public water system.
17. There should be a reliable and adequate electricity supply and emergency lighting to permit safe exit. A stand-by generator is desirable for the support of essential equipment, such as incubators,

## Biological Risk

biological safety cabinets, freezers, etc., and for the ventilation of animal cages.

18. There should be a reliable and adequate supply of gas. Good maintenance of the installation is mandatory.
19. Laboratories and animal cages are occasionally the targets of vandals. Physical and fire security must be considered. Strong doors, screened windows and restricted issue of keys are compulsory. Other measures should be considered and applied, as appropriate, to augment security.

The biosafety officer can assist in training and with the development of training aids and documentation. Staff training should always include information on safe methods for highly hazardous procedures that are commonly encountered by all laboratory personnel and which involve:

1. Inhalation risks (i.e., aerosol production) when using loops, streaking agar plates, pipetting, making smears, opening cultures, taking blood/serum samples, centrifuging, etc. [7].
2. Ingestion risks when handling specimens, smears and cultures.
3. Risks of percutaneous exposures when using syringes and needles.
4. Bites and scratches when handling animals.
5. Handling of blood and other potentially hazardous pathological materials.
6. Decontamination and disposal of infectious material.

---

**Basic biosafety levels 1- 2 Procedures**

✓ Pipetting by mouth must be strictly forbidden.
✓ Materials must not be placed in the mouth. Labels must not be licked.
✓ All technical procedures should be performed in a way that minimizes the formation of aerosols and droplets.
✓ The use of hypodermic needles and syringes should be limited. They must not be used as substitutes for pipetting devices or for any purpose other than parenteral injection or aspiration of fluids from laboratory animals.
✓ All spills, accidents and overt or potential exposures to infectious materials must be reported to the laboratory supervisor. A written record of such accidents and incidents should be maintained.
✓ A written procedure for the clean-up of all spills must be developed and followed.
✓ Contaminated liquids must be decontaminated (chemically or physically) before discharge to the sanitary sewer. An effluent treatment system may be required, depending on the risk assessment for the agent(s) being handled.
✓ Written documents that are expected to be removed from the laboratory need to be protected from contamination while in the laboratory.

---

Figure 19. Basic standard practices for a safety control of laboratory biological.

## 4.4.1.2. Biosafety Levels 3 and 4

The laboratory design and facilities for basic laboratories – Biosafety Levels 1 and 2 apply except where modified as follows:

1. The laboratory must be separated from the areas open to unrestricted traffic flow within the building. Additional separation can be achieved by placing the laboratory at the blind end of a corridor, or by constructing a partition and door or access through an anteroom (e.g., a double-door entry or basic laboratory – Biosafety Level 2), which describes a specific area designed to maintain the pressure differential between the laboratory and its adjacent space. The anteroom should have facilities for separating clean and dirty clothing and a shower may also be necessary.
2. The anteroom doors may be self-closing and interlocking so that only one door is open at a time. A break-through panel may be provided for emergency exit use.
3. Surfaces of walls, floors and ceilings should be water-resistant and easy to clean. Openings through these surfaces (e.g., for service pipes) should be sealed to facilitate decontamination of the room(s).
4. The laboratory room must be sealable for decontamination. Air-ducting systems must be constructed to permit gaseous decontamination.
5. The windows must be closed, sealed and break-resistant.
6. A hand-washing station with hands-free controls should be provided near each exit door.
7. A controlled ventilation system must be provided to maintain a directional airflow into the laboratory room. A visual monitoring device with or without alarm(s) should be installed, so that staff can at all times ensure that proper directional airflow into the laboratory room is maintained.
8. The building ventilation system must be so constructed that air from the containment laboratory – Biosafety Level 3 is not recirculated to other areas within the building. Air may be high-efficiency particulate air (HEPA) filtered, reconditioned and recirculated

within that laboratory. When exhaust air from the laboratory (other than from biological safety cabinets) is discharged to the outside of the building, it must be dispersed away from occupied buildings and air intakes. Depending on the agents in use, this air may be discharged through HEPA filters. A heating, ventilation and air-conditioning (HVAC) control system may be installed to prevent sustained positive pressurization of the laboratory. Consideration should be given to the installation of audible or clearly visible alarms to notify personnel of HVAC system failure.

9. All HEPA filters must be installed in a manner that permits gaseous decontamination and testing.

10. Biological safety cabinets should be sited away from walking areas and out of crosscurrents from doors and ventilation systems.

11. The exhaust air from Class I or Class II biological safety cabinets (see Chapter 6), which will have been passed through HEPA filters, must be discharged in such a way as to avoid interference with the air balance of the cabinet or the building exhaust system.

12. An autoclave for the decontamination of contaminated waste material should be available in the containment laboratory. If infectious waste has to be removed from the containment laboratory for decontamination and disposal, it must be transported in sealed, unbreakable and leakproof containers according to national or international regulations, as appropriate.

13. Backflow-precaution devices must be fitted to the water supply. Vacuum lines should be protected with liquid disinfectant traps and HEPA filters, or their equivalent. Alternative vacuum pumps should also be properly protected with traps and filters.

14. The containment laboratory – Biosafety Level 3 facility design and operational procedures should be documented. Figure 20 shows an example of laboratory design for Biosafety Level 3.

The maximum containment laboratory – Biosafety Level 4 is designed for work with Risk Group 4 microorganisms. Before such a laboratory is constructed and put into operation, intensive consultations should be held with institutions that have had experience of operating a similar facility. Operational maximum containment laboratories – Biosafety Level 4 should be under the control of national or other appropriate health authorities.

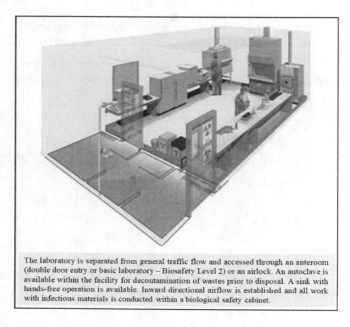

The laboratory is separated from general traffic flow and accessed through an anteroom (double door entry or basic laboratory – Biosafety Level 2) or an airlock. An autoclave is available within the facility for decontamination of wastes prior to disposal. A sink with hands-free operation is available. Inward directional airflow is established and all work with infectious materials is conducted within a biological safety cabinet.

Figure 20. Biosafety Level 3 laboratory (provided by CUH2A, Princeton, NJ, USA).

# REFERENCES

[1] Phillips, G.B. (1969). Control of Microbiological Hazards in the Laboratory. *American Industrial Hygiene Association Journal* 30:170-176. [3, I, p. 23].

[2] Fleming, D., Richardson, J., Tulis, J., Vesley, D. (1995). *Laboratory Safety – Principles and Practices*, page 212, ASM Press Washington, D.C.

[3] Miller, C.D., Songer, J.R., and Sullivan, J.F. (1987). *A Twenty-Five Year Review of Laboratory-Acquired Human Infections at the National Animal Disease Center*. American Industrial Hygiene Association 48:271-275. [2, C, p. 10], [3, B, p. 13], [3, F, p. 18].

[4] Richardson, J.H, editor; and Barkley, W.E, eds. (1984). *Biosafety in Microbiological and Biomedical Laboratories*. U.S. Public Health Service. Centers for Disease Control and National Institutes of Health. HHS Publication No. (CDC) 84-8395. Washington, DC: U.S. Government Printing Office. [2, B, p. 9], [2, C, p. 12], [3, A, p. 13], [3, B, p. 13], [3, G, p. 21], [3, K, pp. 30], [5, C, p. 54], [5, D, pp. 59, 62], [Appendix B].

[5] Kent, P.S. and Kubica, G.P. (1985). *A Guide to the Level III Laboratory*. U.S. Department of Health and Human Services, U.S. Public Health Service, Centers for Disease Control. Atlanta, GA: Centers for Disease Control. [3, J, p. 26].

[6] World Health Organization. 2004. *Laboratory Biosafety Manual–* 3rd ed. ISBN 92 4 154650 6. Geneva: WHO.

[7] Hanel, E., and Halbert, M.M. (1986). *Pipetting. Pp. 204-214 in Laboratory Safety: Principles and Practices*, BM Miller, editor; DHM Groschel, editor; J.H Richardson, editor; D Vesley, editor; JR Songer, editor; , RD Housewright, editor; and WE Barkley, editors. Washington, DC: American Society for Microbiology. [3, I, p. 23], [3, J, p. 26].

*Chapter 5*

# RADIOLOGICAL SAFETY

Laboratory personnel may be exposed to ionizing radiation with the use of isotopes and tracers, and non-ionizing radiation emitted by instruments (microwave ovens), light sources, lasers, and UV lamps [1]. The use of radionuclides must be carried out in well-identified and confined areas, within which appropriate protective measures must be taken to limit irradiation. With non-ionizing radiation, the possible risks can be tolerated by sight, in particular by the use of laser devices without the use of special eye shields, and for prolonged exposure to ultraviolet light [2].

## 5.1. IONIZING RADIATIONS

Numerous radioactive substances are used in the chemical, biological, biomedical and biotechnological laboratory of research and analysis. The most common are:

- Carbon-14 ($^{14}$C), used in research to ensure that new potential drugs are metabolised without the formation of harmful products.
- Trizio ($^3$H), used for biomedicine studies and drug metabolism to ensure the safety of potential new drugs.

- Iodine-125 ($^{125}$I), used in vitro for diagnostic kits.
- Phosphorus-32 ($^{32}$P), used in molecular biology and in genetic research [3].
- Phosphorus-33 ($^{33}$P) and Sulfur-35 ($^{35}$S), often replace the ($^{32}$P) for safety reasons because their Beta emission is low energy.

Those who procure, use, possess, transports, transfers, or disposes of regulated radioactive materials or radiation generating devices must:

- Notify, in writing, the Designated Authority of the nature of the material or device and provide a description of the intended use, the location of use and storage, and all transportation and disposal requirements.
- Ensure appropriate authorization or a permit if a licensed or regulated radiological device or radioactive material is to be used on Reclamation property.

## 5.1.1. Qualified Personnel and Occupational Dose Limits

Operations involving the risk of ionizing radiation or the use of radioactive material or radiation generators must be carried out under the direct supervision of a person, designated in writing by the Radiation Safety Officer (RSO), who is qualified and responsible for radiological safety [4]. This person will conduct surveys and assess any specialized assistance needed to ensure compliance with radiation protection standards. The RSO must be technically qualified and meet the following experience, training, and education requirements:

1. Formally trained in radiation protection, including radiation physics; interaction of the radiation with the matter; mathematics necessary for the subject matter; biological effects of radiation; type and use of instruments for the detection, monitoring, and surveying radiation; radiation safety techniques and procedures; and use of

## Radiological Safety

time, distance, shielding, engineering controls, and PPE to reduce radiation exposure.

2. Pratical training regarding all equipment, instrumentation, procedures, and theory used.
3. Knowledge of regulations (NRC, EPA, Department of Energy (DOE), DOT, and DOI) related to radioactive materials, radiation generating devices and radioactive and mixed waste.
4. Knowledge of standards and conservation requirements for work with radioactive materials and radiation generating devices.

In order to limit the harmful effects of ionizing radiation, the use of radioisotopes must be monitored and must conform to the relevant national standards [5]. Radiation protection is managed on the basis of four principles:

1. Minimize radiation exposure time.
2. Maximize the distance from the radiation source.
3. Shielding of the radiation source.
4. Substition of the use of radionuclides with non-radiometric techniques.

Security tasks include the following.

### Time

The time of exposure experienced during manipulations of radioactive material can be reduced by:

- Practising new and unfamiliar techniques without using the radionuclide until the techniques are mastered.
- Working with radionuclides in a deliberate and timely manner without rushing.
- Make sure that all radioactive sources are returned to storage immediately after use.

- Removal of radioactive waste from the laboratory at frequent intervals.
- Spend as little time as possible in the radiation area or in the laboratory.
- Effective time management exercise and scheduling of laboratory manipulations involving radioactive material.

Less time spent in a radiation field, the lower the personal dose received, as described in the equation:

Dose = Dose rate x time

### *Distance*

The dose rate for most $\gamma$- and X-radiation varies as the inverse square of the distance from a point source:

Dose rate = Constant /Distance$^2$

Doubling the distance from a radiation source will result in reducing the exposure by one-fourt over the same period of time. Various devices and mechanical aids are used to increase the distance between the operator and the radiation source, e.g., long-handled tongs, forceps, clamps and remote pipetting aids. Note that a small increase in distance can result in significant decrease in the dose rate.

### *Shielding*

Radiation energy-absorbing or attenuating shields placed between the source and the operator or other occupants of the laboratory will help limit their exposure. The choice and thickness of any shielding material depends on the penetrating ability (type and energy) of the radiation. A barrier of acrylic, wood or lightweight metal, thickness 1.3–1.5 cm, provides shielding against high-energy $\beta$ particles, whereas high-density lead is needed to shield against high energy $\gamma$ and X-radiation.

## Substitution

Radionuclide-based materials should not be used when other techniques are available. If substitution is not possible, the radionuclide must be used with the least penetrating power or energy.

Operations involving the dangers of regulated radiation and users of radioactive material or radiation generators must develop and implement a radiation safety program. The RSO must manage the program and base it on the principles of radiation safety ALARA (As Low As Reasonably Acceptable). Review the program at least once a year.

Instruct all staff entering an area where radioactive material or radiation generating devices are used and where there is a potential for an individual to receive a Total Effective Dose Equivalent (TEDE) of 100 mrem or more in one year in:

- the presence of the material or device;
- health and safety problems associated with radiationexposure, including potential radiation effects on a pregnantfemale, fetus or embryo;
- precautions and controls used to control exposure;
- correct use of instrumentation and dosimetry in the area;
- the radiation Safety programme;
- their rights and responsibilities.

No employee under 18 years of age will perform work with or around ionizing radiation. The dose to an embryo/fetus must not exceed the monthly equivalent dose of 0.05 rem during the entire gestation period.

The annual limit for occupational workers is the more limiting of:

a. The total effective dose equivalent = 5,000 millirem (0.05 Sv); or
b. The sum of the deep dose equivalent and the committed dose equivalent to any individual organ or tissue (other than the lens of the eye) = 50,000 millirem (0.5 Sv). Table 18 shows the annual limits to the lens of the eye, to the skin, and to the extremities.

## Table 18. Exposure to ionizing radiation

| Body part | Annual limits (NRC)[*] | Suggested ALARA limits[**] |
|---|---|---|
| Whole body | 5 rem (50 mSv {millisievert})[***] | 0.1 rem (1 mSv) |
| Lens of eye | 15 rem (150 mSv) | 0.15 rem (1.5 mSv) |
| Skin | 50 rem (500 mSv) | 0.5 rem (5 mSv) |
| Hands/feet | 50 rem (500 mSv) | 0.5 rem (5 mSv) |

[*]An annual limit which is the more limiting of: 5 rems TEDE, 15 rems to the lens of the eye, or 50 rems shallow dose equivalent to the skin or any extremity.

[**]To keep doses ALARA, the user will set administrative action levels below the annual dose limits. These action levels must be realistic and attainable. Suggested action levels are the more limiting of: 0.1 rems TEDE, 0.15 rems to the lens of the eye, or 0.5 rems shallow dose equivalent to the skin or any extremity.

[***]10 mSv = 1 rem.

The dose limit to an embryo/fetus during the entire pregnancy due to occupational exposure of a declared pregnant woman is 500 millirem (5 mSv).

Care shall be taken so that no more than 50 millirem (0.5 mSv) be received during any one month during a declared pregnancy. Efforts shall be made to avoid substantial variation above the uniform monthly exposure rate to a declared pregnant woman. If the pregnant woman has not notified DRS of her estimated date of conception, the dose to the fetus shall not exceed 50 millirem (0.5 mSv) per month during the remainder of the pregnancy. If, at the time the pregnant woman informs DRS of the estimated date of conception, the dose for the embryo/fetus has exceeded 450 millirem (4.5 mSv), the limit for the remainder of the pregnancy is 50 millirem (0.5 mSv).

## 5.1.2. Radiation Monitoring and Dosimetry

Users of radioactive material or radiation generating devices must conduct surveys and monitoring to ensure occupational dose limits are not exceeded.

Swipte-test each sealed source, other than those exempt by size or specific regulation, for leakage at not greater than 6-month intervals and maintain records for each test. If the sample indicated a contamination activity greater than 0.005 microcuries (µCi), withdraw the source from use and notify the RSO immediately.

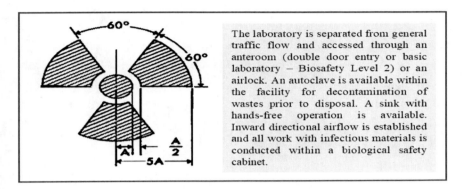

Figure 21. International symbol of radiation.

Use tools for monitoring and measuring the radiation appropriate for the type and intensity of the radiation analysed, calibrated on a traceable source and controlled at an operational level before each use. Users of radioactive material or radiation generating devices and visitors or personnel performing work tasks in the area must coordinate with the RSO for appropriate use of dosimetry whenever one of the following situations exist:

1. an individual enters a Radiation Area (>5 mrem in any 1 hour);
2. an individual has the potential to receive greater than 0.5 rem in 1 year.

All individuals must wear personnel monitoring equipment within the radiation areas as defined above. Supervisors are responsible for ensuring compliance. It processes all external dosimetry at a National Voluntary Laboratory Accreditation Program (NVLAP) certified laboratory. Users of sources of unsealed radioactive material or personnel working on a site of radioactive hazardous waste must estabilish a RSO-approved internal dosimetry programme in the event of a potential for a worker to receive a dose internal higher than 0.5 rem per year.

## 5.1.3. Warning, Postings and Containers' Labeling Requirements

Containers of RM include laboratory waste pails and drums and some vessels used in the research protocols. All shall have a label affixed to the container that displays the radiation symbol and the word *"Caution, Radioactive Material"* or *"Danger, Radioactive Material"*. In addition, the label must contain sufficient information about its contents, allowing people who manipulate the container or who work nearby to take the necessary precautions and minimize radiation exposures.

Labels are not mandatory when:

- The container is ready for transport and labelled according to State and Federal regulations; it is in a protected area under the control of the RSO and a description of the isotopic content is in the room; or
- Laboratory containers keep the materials in transitional procedures that last only a few hours and are disposed of immediately after an experiment and under the control of or in the presence of the Authorized User.

**Table 19. The four warning symbols of the radiation safety warning**

| CAUTION RADIATION AREA | Areas where the radiation field is equal to or greater than 5 mrem (0.05 mSv) in any one hour and less than 100 mrem (1 mSv) in any 1 hour | CAUTION HIGH RADIATION AREA | Areas where radiation field is equal to or greater than 100 mrem in any one hour (0.1 mSv) and less than 500 rads in any 1 hour |
|---|---|---|---|
| CAUTION AIRBORNE RADIATION AREA | Areas where airborne radioactive material concentrations are greater than the derived air concentration (DAC) limits listed in 10 CFR 20 appendix B*. | CAUTION RADIOACTIVE MATERIALS | Rooms where quantities of radioactive materials in excess of 10 times the 10 CFR 20 appendix C** quantities are used or stored. |

*https://www.nrc.gov/reading-rm/doc-collections/cfr/part020/appb/
**https://www.nrc.gov/reading-rm/doc-collections/cfr/part020/part020-appc.html

## Radiological Safety

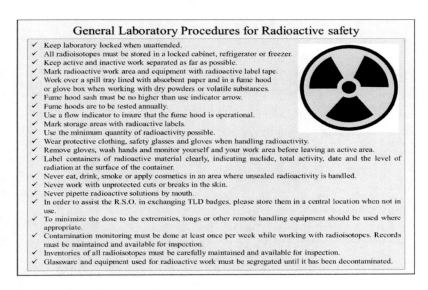

Figure 22. Main roules for Laboratory radioactive safety.

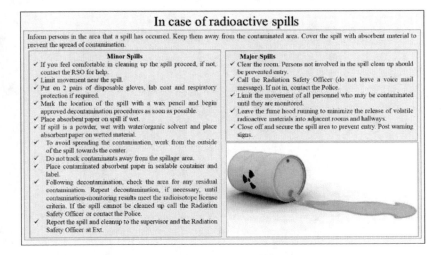

Figure 23. How to correclty remove radioactive material from the contaminated area.

Users who receive or expect to receive a package containing radioactive material must follow the package receipt procedures listed in 10 CFR 20.1906, Procedures for Receiving and Opening Packages [6]. Finally, the RSO must post an NRC Form 3 "Notice to Employees" in a location visible to all employees who work with or around radioactive materials.

## 5.1.4. Radioactive Waste Disposal

Radioactive sealed sources (and gauges), when no longer needed, may be returned (transferred) to the manufacturer. Notify the local RSO and amend or terminate any applicable licenses or permits. Dispose of radioactive waste appropriately, in accordance with Federal, State, and local regulations, only after coordinating with the designated RSO.

## 5.1.5. Safety Data Sheets

### $^3$H - Hydrogen-3 [Tritium]

**I. PHYSICAL DATA**
Radiation: Beta (100% abundance)
Energy: Max.: 18.6 keV; Average: 5.7 keV
Half-Life [T½] : Physical T½: 12.3 years
Biological T½: 10 - 12 days
Effective T½: 10 - 12 days*
* Large liquid intake (3-4 liters/day) reduces effective T½ by a factor of 2+; $^3$H is easily flushed from the body
Specific Activity: 9650 Ci/g [357 TBq/g] max.
Beta Range: Air: 6 mm [0.6 cm; 0.25 inches]
Water: 0.006 mm [0.0006 cm; 3/10,000 inches]
Solids/Tissue: Insignificant [No 3H betas pass through the dead layer of skin]

**II. RADIOLOGICAL DATA**
Radiotoxicity: Least radiotoxic of all nuclides; CEDE, ingestion or inhalation:
Tritiated water: 1.73E-11 Sv/Bq (0.064 mrem/uCi) of $^3$H intake
Organic Compounds: 4.2E-11 Sv/Bq (0.16 mrem/uCi) of $^3$H intake
Critical Organ: Body water or tissue
Exposure Routes: Ingestion, inhalation, puncture, wound, skin contamination absorption
Radiological Hazard: External Exposure - None from weak $^3$H beta
Internal Exposure and Contamination - Primary concern

**III. SHIELDING**
None required - not an external radiation hazard

**IV. DOSIMETRY MONITORING**
Urine bioassay is the only readily available method to assess intake [for tritium, no intake = no dose] Be sure to provide a urine sample to Radiation Safety for confirmatory bioassay whenever your annual $^3$H use exceeds 8 mCi. If negative, no further bioassay is required unless use exceeds 100 mCi at one time or 1000 mCi in one year, or after any accident/incident in which an intake is suspected

**V. DETECTION AND MEASUREMENT**
Liquid Scintillation Counting is the only readily available method for detecting $^3$H
NOTE: PORTABLE SURVEY METERS WILL NOT DETECT LABORATORY QUANTITIES OF $^3$H

**VI. SPECIAL PRECAUTIONS**
- Avoid skin contamination [absorption], ingestion, inhalation, and injection [all routes of intake]
- Many tritium compounds readily penetrate gloves and skin; handle such compounds remotely and wear double gloves, changing the outer pair at least every 20 minutes.
- While tritiated DNA precursors are considered more toxic that $^3$H$_2$O, they are generally less volatile and hence do not normally present a greater hazard.
- The inability of direct-reading instruments to detect tritium and the slight permeability of most material to [tritiated] water & hydrogen [tritium] facilitates undetected spread of contamination. Use extreme care in handling and storage [e.g. sealed double or multiple containment] to avoid contamination, especially with high specific activity compounds.

# $^{14}$C - Carbon-14

### I. PHYSICAL DATA
Radiation: Beta (100% abundance)
Energy: Max.: 156 keV; Average: 49 keV
Half-Life [T½] : Physical T½: 5730 years
Biological T½: 12 days
Effective T½: Bound - 12 days; unbound - 40 days
Specific Activity: 4.46 Ci/g [0.165 TBq/g] max.
Beta Range: Air: 24 cm [10 inches]
Water/Tissue: 0.28 mm [0.012 inches]
[~1% of 14C betas transmitted through dead skin layer, i.e. 0.007 cm depth]
Plastic: 0.25 mm [0.010 inches]

### II. RADIOLOGICAL DATA
Radiotoxicity: 0.023 mrem/uCi of $^{14}CO_2$ inhaled;
2.09 mrem/uCi organic compounds inhaled/ingested
Critical Organ: Fat tissue [most labeled compounds]; bone [some labeled carbonates]
Exposure Routes: Ingestion, inhalation, puncture, wound, skin contamination absorption
Radiological Hazard: External Exposure – None from weak 14C beta
Internal Exposure & Contamination - Primary concern

### III. SHIELDING
None required - mCi quantities not an external radiation hazard

### IV. DOSIMETRY MONITORING
Urine bioassay is the most readily available method to assess intake [for $^{14}$C, no intake = no dose]
Provide a urine sample to Radiation Safety after any accident/incident in which an intake is suspected

### V. DETECTION AND MEASUREMENT
Portable Survey Meters: Geiger-Mueller [~10% efficiency];
Beta Scintillator [~5% efficiency]
Wipe Test: Liquid Scintillation Counting is the best readily available method for counting $^{14}$C wipe tests

### VI. SPECIAL PRECAUTIONS
> Avoid skin contamination [absorption], ingestion, inhalation, & injection [all routes of intake]
> Many $^{14}$C compounds readily penetrate gloves and skin; handle such compounds remotely and wear double gloves, changing the outer pair at least every 20 minutes.

# ³²P - Phosphorous-32

### I. PHYSICAL DATA
Radiation: Beta (100% abundance)
Energy: Maximum: 1,710 keV; Average: 695 keV
Half-Life [T½] : Physical T½: 14.29 days
Biological T½: Bone ~ 1155 days; Whole Body ~ 257 days[1]
Effective T½: 14.29 days
Specific Activity: 286,500 Ci/g [10,600 TBq/g] max.
Beta Range: Air: 610 cm [240 inches; 20 feet]
Water/Tissue: 0.76 cm [0.33 inches]
Plastic: 0.61 mm [3/8 inches]

### II. RADIOLOGICAL DATA
Radiotoxicity[2]: 94.7 mrem/uCi [Lung] & 15.5 mrem/uCi [CEDE] of ³²P inhaled
29.9 mrem/uCi [Bone Marrow] and 8.77 mrem/uCi [CEDE] of ³²P ingested
Critical Organ: Bone [soluble ³²P]; Lung [Inhalation]; GI Tract [Ingestion - insoluble compounds]
Exposure Routes: Ingestion, inhalation, puncture, wound, skin contamination absorption
Radiological Hazard: External Exposure [unshielded dose rate at 1 mCi ³²P vial mouth[3]: approx. 26 rem/hr], Internal Exposure and Contamination

### III. SHIELDING
Shield ³²P with 3/8 inch Plexiglas and monitor for Bremstrahlung; If Bremstrahlung X-rays detected outside Plexiglas, apply 1/8 to 1/4 inch lead [Pb] shielding outside Plexiglas
The accessible dose rate should be background but must be < 2 mR/hr

### IV. DOSIMETRY MONITORING
Wear radiation dosimetry monitoring badges [body & ring] if regularly handling mCi quantities of ³²P

### V. DETECTION AND MEASUREMENT
Portable Survey Meters: Geiger-Mueller
Wipe Test: Liquid Scintillation Counting is an acceptable method for counting ³²P wipe tests

### VI. SPECIAL PRECAUTIONS
- Avoid skin contamination [absorption], ingestion, inhalation, and injection [all routes of intake].
- Store ³²P (including waste) behind Plexiglas shielding [3/8 inch thick]; survey (with GM meter) to check adequacy of shielding (accessible dose rate < 2 mR/hr; should be background); apply lead [Pb] shielding outside Plexiglas if needed.
- Use 3/8 inch Plexiglas shielding to minimize exposure while handling ³²P.
- Use tools [e.g. Beta Blocks] to handle ³²P sources and contaminated objects; avoid direct hand contact.
- Always have a portable survey meter present and turned on when handling ³²P.
- 32P is not volatile, even when heated, and can be ignored as an airborne contaminant[4] unless aerosolized.

1 NCRP Report No. 65, p.88
2 Federal Guidance Report No. 11 [Oak Ridge, TN; Oak Ridge National Laboratory, 1988], p. 122, 156
3 Dupont/NEN, Phosphorous-32 Handling Precautions [Boston, MA; NEN Products, 1985]
4 Bevelacqua, J. Contemporary Health Physics [New York; John Wiley & Sons, 1995], p. 282

# $^{35}$S - Sulfur-35

### I. PHYSICAL DATA
Radiation: Beta (100% abundance)
Energy: Maximum: 167.47 keV; Average: 48.8 keV
Half-Life [T½] : Physical T½: 87.44 days
Biological T½: 623 days [unbound 35S]; 90 days [bound 35S]
Effective T½: 44 - 76 days [unbound 35S]
Specific Activity: 42,707 Ci/g [1,580 TBq/g] max.
Beta Range: Air: 26 cm [10.2 inches]
Water/Tissue: 0.32 mm [0.015 inches]
Plastic: 0.25 mm [0.010 inches]

### II. RADIOLOGICAL DATA
Radiotoxicity[1]: 2.48 mrem/uCi [CEDE] of 35S inhaled
0.733 mrem/uCi of $^{35}$S ingested
Critical Organ: Testis
Exposure Routes: Ingestion, inhalation, puncture, wound, skin contamination absorption
Radiological Hazard: External Exposure – None from weak $^{35}$S beta
Internal Exposure and Contamination - Primary concern

### III. SHIELDING
None required - mCi quantities not an external radiation hazard

### IV. DOSIMETRY MONITORING
Urine bioassay is the most readily available method to assess intake [for 35S, no intake = no dose]
Provide a urine sample to Radiation Safety after any accident/incident in which an intake is suspected

### V. DETECTION AND MEASUREMENT
Portable Survey Meters: Geiger-Mueller [~10% efficiency]
Beta Scintillator [~5% efficiency]
Wipe Test: Liquid Scintillation Counting is the best readily available method for counting 35S wipe tests

### VI. SPECIAL PRECAUTIONS
- Avoid skin contamination [absorption], ingestion, inhalation, & injection [all routes of intake]
- Many 35S compounds and metabolites are slightly volatile and may create contamination problems if not sealed or otherwise controlled. This occurs particularly when 35S amino acids are thawed, and when they are added to cell culture media and incubated. Therefore vent thawing 35S vials in a hood. Incubators used with 35S will have an activated charcoal trap placed in the incubator. Possibility of volatilization must be taken into account when surveying after use.

1 Federal Guidance Report No. 11 [Oak Ridge, TN; Oak Ridge National Laboratory, 1988], p. 122,156

# $^{45}$C – Calcium - 45

### I. PHYSICAL DATA
Radiation: Beta (100% abundance)
Energy: Maximum: 257 keV; Average: 77 keV
Half-Life [T½] : Physical T½: 162.61 days
Biological T½: Bone ~ 18,000 days[1]
Effective T½: 163 Days
Specific Activity: 17,800 Ci/g [659 TBq/g] max.
Beta Range: Air: 52 cm [20 inches]
Water/Tissue: 0.062 cm [0.024 inches]
Plastic
(Lucite): 0.053 cm [0.021 inches]

### II. RADIOLOGICAL DATA
Radiotoxicity[2]: 35.8 mrem/uCi [Lung] and 16.2 mrem/uCi [Bone] of $^{45}$Ca inhaled
19.4 mrem/uCi [Bone] and 3.2 mrem/uCi [CEDE] of $^{45}$Ca ingested
Critical Organ: Bone; Lung [Inhalation]
Exposure Routes: Ingestion, inhalation, puncture, wound, skin contamination absorption
Radiological Hazard: External Exposure - mCi quantities not considered an external hazard
Internal Exposure and Contamination - Primary concern

### III. SHIELDING
None required - mCi quantities not an external radiation hazard

### IV. DOSIMETRY MONITORING
Urine bioassay is the most readily available method to assess intake. Provide a urine sample to Radiation Safety after any accident/incident in which an intake is suspected.
No dosimetry badges needed to work with mCi quantities of $^{45}$Ca.

### V. DETECTION AND MEASUREMENT
Portable Survey Meters: Geiger-Mueller
Wipe Test: Liquid Scintillation Counting works well for counting $^{45}$Ca wipe tests

### VI. SPECIAL PRECAUTIONS
➢ Avoid skin contamination [absorption], ingestion, inhalation, and injection [all routes of intake]

---

1 "Calcium-45 Handling Precautions", E.I. DuPont de Numours and Co., NEN Products [Boston, MA; 1985]
2 Federal Guidance Report No. 11 [Oak Ridge, TN; Oak Ridge National Laboratory, 1988], p. 122, 156

# $^{125}$I - Iodine - 125

**I. PHYSICAL DATA**
Radiation: Gamma - 35.5 keV (7% abundance)
X-ray - 27 keV (113% abundance)
Gamma Constant: 0.27 mR/hr per mCi @ 1.0 meter [7.432E-5 mSv/hr per MBq @ 1.0 meter][1]
Half-Life [T½] : Physical T½: 60.14 days
Biological T½: 120-138 days (unbound iodine)
Effective T½: 42 days (unbound iodine)
Specific Activity: 1.73E4 Ci/g [642 TBq/g] max.

**II. RADIOLOGICAL DATA**
Radiotoxicity[2]: 3.44E-7 Sv/Bq (1273 mrem/uCi) of $^{125}$I ingested [Thyroid]
2.16 E-7 Sv/Bq (799 mrem/uCi) of $^{125}$I inhaled [Thyroid]
Critical Organ: Thyroid Gland
Intake Routes: Ingestion, inhalation, puncture, wound, skin contamination (absorption);
Radiological Hazard: External & Internal Exposure; Contamination

**III. SHIELDING**
Lead [Pb]
Half Value Layer [HVL]: 0.02 mm (0.0008 inches)
Tenth Value Layer [TVL]: 0.07 mm (0.003 inches)
The accessible dose rate should be background but must be < 2 mR/hr

**IV. DOSIMETRY MONITORING**
Always wear radiation dosimetry monitoring badges [body & ring] whenever handling > 10 mCi of $^{125}$I
Conduct a baseline thyroid scan prior to first use of 1 mCi or more of radioactive iodine
Conduct thyroid scan no earlier than 6 hours but within 72 hours of handling 1 mCi or more of $^{125}$I or after any suspected intake

**V. DETECTION AND MEASUREMENT**
Portable Survey Meters:
Geiger-Mueller
Low Energy Gamma Detector [~19% eff. for $^{125}$I] for contamination surveys
Wipe Test: Liquid Scintillation Counter or Gamma Counter

**VI. SPECIAL PRECAUTIONS**
➢ Avoid skin contamination [absorption], ingestion, inhalation, and injection [all routes of intake]
➢ Use shielding [lead or leaded Plexiglas] to minimize exposure while handling mCi quantities of $^{125}$I
➢ Avoid making low pH [acidic] solutions containing $^{125}$I to avoid volatilization
➢ For Iodinations:
- Use a cannula adapter needle to vent stock vials of $^{125}$I used; this prevents puff releases
- Cover test tubes used to count or separate fractions from iodinations with parafilm or other tight caps to prevent release while counting or moving outside the fume hood.

1 Health Physics & Radiological Health Handbook, 3rd Ed. [Baltimore, MD; Williams & Wilkins, 1998] p. 6-11
2 Federal Guidance Report No. 11 (Oak Ridge TN; Oak Ridge National Laboratory, 1988) P. 136, 166

## 5.2. NON-IONIZING RADIATIONS

Employers will use qualified, competent persons and appropriately calibrated monitoring equipment to assess, survey, and evaluate non-ionizing radiation and field strengths, employee exposures, and control measures.

Comply with the manufacturer's requirements and restrictions in accordance with current ANSI Z136.1, American National Standard for the Safe Use of Lasers when installing and using lasers and laser systems.

Improperly used laser devices are potentially dangerous. Effects can range from mild skin burns to irreversible injury to the skin and eye. The biological damage caused by lasers is produced through thermal, acoustical and photochemical processes. Lasers and laser systems are assigned one of four broad Classes, depending on the potential for causing biological damage.

*Class I:* cannot emit laser radiation at known hazard levels (typically continuous wave: cw 0.4 µW at visible wavelengths). Users of Class I laser products are generally exempt from radiation hazard controls during operation and maintenance (but not necessarily during service). Since lasers are not classified on beam access during service, most Class I industrial lasers will consist of a higher class (high power) laser enclosed in a properly interlocked and labeled protective enclosure. In some cases, the enclosure may be a room (walk-in protective housing) which requires a means to prevent operation when operators are inside the room.

- *Class IA:* a special designation that is based upon a 1000-second exposure and applies only to lasers that are "not intended for viewing" such as a supermarket laser scanner. The upper power limit of Class I.A. is 4.0 mW. The emission from a Class I.A. laser is defined such that the emission does not exceed the Class I limit for an emission duration of 1000 seconds.

- *Class II:* low-power visible lasers that emit above Class I levels but at a radiant power not above 1 mW. The concept is that the human aversion reaction to bright light will protect a person. Only limited controls are specified.

- *Class IIIA:* intermediate power lasers (cw: 1-5 mW). Only hazardous for intrabeam viewing. Some limited controls are usually recommended.

# Radiological Safety

- *Class IIIB:* moderate power lasers (cw: 5-500 mW, pulsed: 10 $J/cm^2$ or the diffuse reflection limit, whichever is lower). In general Class IIIB lasers will not be a fire hazard, nor are they generally capable of producing a hazardous diffuse reflection. Specific controls are recommended.
- *Class IV:* High power lasers (cw: 500 mW, pulsed: 10 $J/cm^2$ or the diffuse reflection limit) are hazardous to view under any condition (directly or diffusely scattered) and are a potential fire hazard and a skin hazard. Significant controls are required of Class IV laser facilities.

**Table 20. Probable biological effects of the hazard classes**

| Laser classifications-summary of hazards | | | | | | |
|---|---|---|---|---|---|---|
| **Class** | **UV** | **VIS** | **NIR** | **IR** | **Direct ocular** | **Diffuse ocular** | **Fire** |
| I | X | X | X | X | No | No | No |
| IA | -- | X* | -- | -- | Only after 1000 sec | No | No |
| II | -- | X | -- | -- | Only after 0.25 sec | No | No |
| IIIA | X | X** | X | X | Yes | No | No |
| IIIB | X | X | X | X | Yes | Only when laser output is near Class IIIB limit of 0.5 Watt | No |
| IV | X | X | X | X | Yes | Yes | Yes |

X = class applies in wavelength range.

*Class IA applicable to lasers "not intended for viewing" *ONLY*.

**CDRH Standard assigns Class IIIA to visible wavelengths *ONLY*. ANSI Z 136.1 assigns Class IIIA to all wavelength ranges.

# REFERENCES

[1] American National Standards Institute, *American National Standard for the Safe Use of Lasers*: ANSI Z 136.1 (1993), Laser Institute of America, New York, NY (1993).

[2] University of Washington. *Radiation Safety Manual* (2015, March). Retrieved from https://www.ehs.washington.edu/manuals/rsmanual/

[3] Castegnaro M., Brésil H., Manin J., P. (1993) *Some Safety Procedures for Handling $^{32}P$ During Postlabelling Assays*. In: Phillips D.H., Castegnaro M., Bartsch H. Postlabelling methods for detection of DNA adducts. IARC scientifique publications N° 124, IARC, Lyon, France.

[4] Zakharyuta, Anastasia and Şen, Canhan and Avaz, Merve Senem and Akkaş, Tuğçe and Pürçüklü, Sibel and Baytekin Birkan, Tuğba and Gönül, Turgay and Yerdelen, Bilge and Cebeci, Fevzi Çakmak and İnce, Adna (2016) *Laboratory Safety Handbook*. Sabanci University, Istanbul. ISBN: 978-605-9178-59-4.

[5] International Commission on Radiological Protection *Recommendation of the International Commission on Radiological Protection* (1990).Annals of ICPR, publication N° 60.

[6] U.S. Nuclear Regulatory Commission, 10 CFR Part 20, *Standards for Protection Against Radiation*, U.S. Nuclear Regulatory Commission, Washington, DC, https://www.nrc.gov/reading-rm/doc-collections/cfr/part020/part020-1906.html

*Chapter 6*

# COLLECTIVE PROTECTION EQUIPMENT

When the risks have been analyzed and it proves impossible to eliminate the risk at its source, protective devices are provided. Protective equipment is put in place to protect personnel from risks that might endanger their health or physical integrity.

Two types of equipment are essential to mitigate business risks and the risks of accidents for workers:

- Collective Protection Equipment (CPE) and
- Personal Protection Equipment (PPE)

## 6.1. CPE GENERALITY

Engineering controls are the most important means of controlling exposure to residual hazards. The engineering controls are all that is built or installed to separate people from chemical, biological or physical hazards, and may include fume hoods, biosafety cabinets, glove boxes, local exhaust ventilation, safety shields and suitable storage facilities.

Hazardous chemicals must be handled and ventilated differently from general laboratory ventilation. Laboratory users should be aware of their chemicals using SDS, such as protecting themselves and laboratory environment from hazardous exposures and taking into account the available engineering controls.

## 6.1.1. Fume Hoods

Fume hoods are used to prevent the release of hazardous and odorous chemical exposure to laboratory, laboratory users and the user. Another substantial reason is to restrict the area affected by splillage inside the hood and the affected air exhausting. The internal air flow through the hood minimizes the leakage of material from the hood.

No large open hood with a low face speed can provide complete safety against all events that may occur in the hood, nor provide protection for volatile airborne contaminants with a threshold limit value (TLV) in the low parts per billion range. For more ordinary exposures, a well-designed hood in a suitably ventilated laboratory can provide adequate protection.

However, some work practices are necessary for the hood to run so capably.

The following working practices are generally required; stricter practices may be necessary in some circumstances. Chapter 2 has already provided some general rules on the correct use of the hood in the analytical laboratory, the following working instructions are more detailed practices and may be necessary in some specific circumstances.

- Keep all apparatus at least 6 inches back from the face of the hood. A stripe on the bench surface is a good reminder.
- Hood sash openings should be kept to a minimum. Hoods are tested (and should be used) with a hood sash opening of 15 inches.
- Do not use the hood as a waste disposal mechanism except for small quantities (<10 mL) of volatile materials.
- Keep the slots in the hood baffle free of obstruction by apparatus or containers.

## Collective Protection Equipment

- Minimize foot traffic past the face of the hood to prevent disruptions in air flow.
- Keep laboratory doors closed when working in the hood.
- Traps, scrubbers or incinerators should be used to prevent toxic and/or noxious materials from being vented into the hood exhaust system.
- Do not place electrical outlets or other sources of spark inside the hood when flammable liquids or gases are present. Permanent electrical outlets are not allowed in the hood.
- Use an appropriate barricade (e.g., a blast shield) if there is a possibility of explosion or implosion.
- Do not remove the door or the hood panels except when necessary to install the appliance; replace the sash or panels before operating.
- Exhaust ports from the hood and supply air vents to the room (Nesbitt units or unit ventilators) should not be blocked.
- Prepare a plan of action in case of an emergency, e.g., a power failure.
- To save energy, turn off the blower and close the hood sash when the hood is not in use.

The annual verification and inspection of the flue gas hood are carried out by the authorized company (manufacturer). The hood inspection program consists of a detailed preliminary inspection followed by standardized annual inspections for all the fume hoods on the campus. This initial inspection will provide basic information, including, but not limited to, the use of the hood, the hood category, the room and building information, and the average capacity of the facial speed.

Follow-up inspections for the appropriate procedure and measurement of face velocity (airflow) will be performed on an annual basis and on request by laboratory users [1]. Upon reaching each inspection, hoods will be categorized with an inspection sticker indicating the speed of the face, the date inspected and the initials of the Inspector. Files including inspection data will be stored in a database.

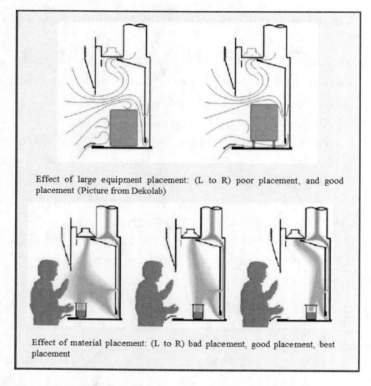

Figure 24. Consequences of positioning of the material under the hood.

A fume hood that has not passed the authorized company inspection and has a warning sign attached, even if the fume hood seems to have the airflow, is not allowed to operate. Laboratory users must organize or coordinate with other available laboratories with operational fume hoods if their work necessitates the usage of a fume hood.

 The ANSI/ASHRAE 2016-110-performance test gives a relative and quantitative determination of the efficiency of the hood containment under a conditioned environment. This method involves releasing a small amount of a tracer gas (either sulfur hexafluoride, or a gas of similar molecular weight and stability) at a fixed rate within the laboratory fume hood, while monitoring the concentration of the tracer material observed in the user's breathing zone using highly specialized testing equipment.

Figure 25. The performance test on hood's efficiency.

## Collective Protection Equipment

The different types of hoods are listed below:

*Standard* – Air volume varies according to the movement of the sash so that as the sash lowered the velocity decreases. Even bypass hoods have similar design but they have an additional vent at the tpo so that as the sash lowered and the sash space is closed, top vent is simultaneously activated, even though the sash opening is getting smaller, the proportion of air volume flowing through the face is smaller and the velocity remains more constant. Although, performance of bypass hoods are better than standard hoods, they are worse than standard hoods in context of energy saving.

*Variable Air Volume* (VAV) – VAV hoods maintain a constant velocity as the sash moves but changes the volume of air. This can be done by a variety of methods including changing motor speed or closing or opening baffles in the duct. Decrease in air usage at lower sash height provides huge amount of energy savings on heating and cooling

*Auxiliary Air* – These hoods additional air-injecting blower at the face of the hood. Auxiliary air type hoods are out-of-use owing to their lower performance than VAV and bypass hoods.

*Ductless hoods* – These hoods are not ducted to outside air but remove contaminants from the air and return it back to the room. HEPA filter, carbon adsorption or catalyst reaction filter may remove the contaminants. Filters should be used in recommended time of use given by the manufacturer. Choosing the appropriate filter according to your contaminant also plays an important role. It is very important that these elements work properly since the air is reticulated and exposure is eminent. Fnless the benefits outweigh the hazards and inconvenience because potential for problems, ductless hoods could pose danger. Also, ductless hoods are not indicated when using many liquid, non-aqueous chemicals since the vapors of these chemicals are heavier than air and ductless hoods do not generally have a rear baffle. As a general rule, the use of hoods without duct is not recommended.

*Clean hoods* – Clean hoods are sometimes called laminar hoods but these are mentioned in Chapter 3 on Biological Risk.

Design of these hoods are based on preventing the work area with HEPA filtered air from contamination. High incidence of air drawn to the hood is filtered and drop gently from the top of into the work area and small percentage (%10) is drawn through the face of the hood. Face velocities of clean hoods are lower than other hoods, however, the hood is designed to capture with high performance in this form. Due to this capture capability, it is important to have a visual capture test (such as a dry ice test) done on these hoods at least annually.

Other capture or contaminant devices include various forms of local exhaust ventilation may have been designed and installed for specific processes, such as

- Elephant Trunk
- Canopy
- Slot and plenum

## 6.1.2. Biological Safety Cabinets

Laboratory techniques can produce aerosols, which may contain hazardous research materials, such as infectious agents that laboratory workers may inhale. Biological safety cabinets (BSC) are a type of primary barrier to contain potentially infectious research materials in order to prevent the exposure of laboratory personnel and contamination of the general environment. Some biological safety cabinets also provide a clean work environment to protect cell cultures or sterile apparatus [2].

*Collective Protection Equipment* 113

There are three classes of biological safety cabinets, designated as Class I, Class II, and Class III. Class I and II cabinets have a protective air barrier across the work opening that separates the laboratory researcher from the work area. Class II cabinets also provide a HEPA-filtered, clean work area to protect the experiment from room contamination. Several variations of Class II cabinets are described below. Class III cabinets have a physical barrier between the operator and the work area. Arm length rubber gloves sealed to glove ports on the cabinet provide the operator with access to the work area. The distinctive features of the three classes of cabinets follow.

*The Class I* cabinet is ventilated for personnel and environmental protection, with an inward airflow away from the operator. It is similar in air movement to a laboratory hood.

- The minimum average face velocity through the work opening is 75 feet per minute (fpm).
- The cabinet exhausts air through a HEPA filter to prevent discharge of most particles to the outside atmosphere.
- This cabinet is suitable for work with low and moderate risk biological agents, where no product protection is required.
- Because of the popularity of Class II cabinets and the product protection they provide, use of Class I cabinets has declined.

*The Class II* cabinet ventilates air for personnel, product, and environmental protection, and has an open front and inward airflow for personnel protection. Product protection comes from HEPA filtered laminar airflow from a diffuser located above the work area. The downflow air splits at the work surface, and exits the work area through grilles located at both the rear and front of the work surface, respectively. The cabinet has HEPA filtered exhausted air for environmental protection. Types of Class II biological safety cabinets are designated A1, A2, B1, and B2.

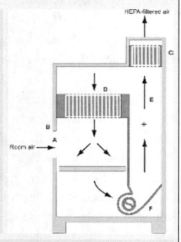

- ✓ The work opening is 8 to 10 inches (20-25 cm) high.
- ✓ The type A cabinet may have a fixed work opening, a sliding sash, or a hinged window.
- ✓ A fan located within the unit provides the intake, recirculated supply air, and the exhaust air.
- ✓ The BSC fan maintains a minimum average inflow velocity of 75 fpm through the work area access opening.
- ✓ Approximately 70% of the cabinet air recirculates through a HEPA filter into the work area from a common plenum, while approximately 30% of the air enters through the front opening and an amount equal to the inflow is exhausted from the cabinet through a HEPA filter.
- ✓ The cabinet may exhaust HEPA filtered air back into the laboratory or exhaust to the environment through an exhaust canopy.
- ✓ The cabinets may have positive pressure contaminated plenums. Contaminated plenums under positive pressure must be gas tight.
- ✓ Type A1 cabinets are suitable for work with low to moderate risk biological agents in the absence of volatile toxic chemicals and volatile radionuclides.

**The Class II, Type A1 BSC** (A) front opening; (B) sash; (C) exhaust HEPA filter; (D) supply HEPA filter; (E) common plenum; (F) blower.

Figure 26. Class II, Type A1 (Formerly Type A) Cabinets.

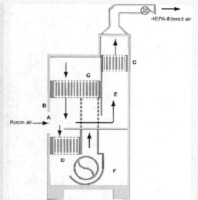

- ✓ The work opening is 8 inches (20 cm) high, with a sliding sash that one can raise for introduction of equipment into the cabinet.
- ✓ Type B1 cabinets have a minimum average inflow velocity of 100 fpm through the work area access opening.
- ✓ The HEPA filtered down flow air is composed largely of uncontaminated recirculated inflow air.
- ✓ Supply fans located in the base of the cabinet, below the work surface, draw air through a grille at the front of the work surface, and supply HEPA filters located directly below the work surface. The fans then force the filtered air through plenums in the sides or the rear of the cabinet and recirculate the air through a diffuser above the work surface. Some cabinets have a secondary supply filter located above the work surface.
- ✓ Approximately 70% of the contaminated down flow air is exhausted through a HEPA filter and a dedicated duct and then discharged outside the building.
- ✓ The remote exhaust fan is generally located on the roof of the building.
- ✓ All biologically contaminated ducts and plenums are under negative pressure or surrounded by negative pressure ducts and plenums. The type B1 cabinet is suitable for work with low to moderate risk biological agents. They are also useful for biological materials treated with minute quantities of toxic chemicals and trace amounts of radionuclides.

**The Class II, Type B1 BSC** (classic design) (A) front opening; (B) sash; (C) exhaust HEPA filter; (D) supply HEPA filter; (E) negative pressure dedicated exhaust plenum; (F) blower; (G) additional HEPA filter for supply air. **Note: The cabinet exhaust needs to be hard connected to the building exhaust system.**

Figure 27. Class II, Type B1 Cabinets.

## Collective Protection Equipment 115

- ✓ The type B2 cabinet has a sliding sash with an 8-inch (20 cm) opening.
- ✓ The type B2 cabinet maintains a minimum average inflow velocity of 100 fpm through the work area access opening.
- ✓ No air recirculates within the cabinet.
- ✓ A supply fan draws air from the laboratory and forces it through a supply HEPA filter located over the work area.
- ✓ A remote exhaust fan, generally located on the roof, pulls all inflow air and supply air through a HEPA filter, and discharges it outside the building. As much as 1200 cubic feet per minute may be exhausted from a 6 ft. cabinet.
- ✓ The cabinet has all contaminated ducts and plenums under negative pressure or surrounded by directly exhausted (not recirculated through work area) negative pressure ducts and plenums.
- ✓ Type B2 cabinets are suitable for work with low to moderate risk biological agents. They are also useful for biological materials treated with toxic chemicals and radionuclides.

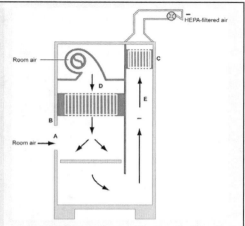

*The Class II, Type B2 BSC* (A) front opening; (B) sash; (C) exhaust HEPA filter; (D) supply HEPA filter; (E) negative pressure exhaust plenum. **Note: The carbon filter in the exhaust system is not shown. The cabinet needs to be hard connected to the building exhaust system.**

Figure 28. Class II, Type B2 ("Total Exhaust") Cabinets.

- ✓ Type A2 cabinets have a minimum of 100 fpm average inflow velocity.
  - o All biologically contaminated ducts and plenums are under negative pressure or surrounded by negative pressure ducts and plenums
  - o They may exhaust HEPA filtered air back into the laboratory or to the environment through and exhaust canopy.
- ✓ Type A2 cabinets are suitable for work with low to moderate risk biological agents. Type A2 cabinets used for work with minute quantities of volatile toxic chemical and trace amounts of radionuclides must be exhausted through properly functioning exhaust canopies. If the cabinet is not ducted, you cannot work with chemicals in the cabinet.
- ✓ The Class II Type A2 BSC is not equivalent to what was formerly called a Class II Type B3 unless it is connected to the laboratory exhaust system.

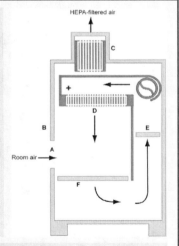

*The tabletop model of a Class II, Type A2 BSC* (A) front opening; (B) sash; (C) exhaust HEPA filter; (D) supply HEPA filter; (E) positive pressure common plenum; (F) negative pressure plenum.. **Note: The A2 BSC should be canopy connected to the exhaust system.**

Figure 29. Class II, Type A2 (Formerly B3) Cabinets.

- ✓ Type A2 cabinets have a minimum of 100 fpm average inflow velocity.
  - o All biologically contaminated ducts and plenums are under negative pressure or surrounded by negative pressure ducts and plenums
  - o They may exhaust HEPA filtered air back into the laboratory or to the environment through and exhaust canopy.
- ✓ Type A2 cabinets are suitable for work with low to moderate risk biological agents. Type A2 cabinets used for work with minute quantities of volatile toxic chemical and trace amounts of radionuclides must be exhausted through properly functioning exhaust canopies. If the cabinet is not ducted, you cannot work with chemicals in the cabinet.
- ✓ The Class II Type A2 BSC is not equivalent to what was formerly called a Class II Type B3 unless it is connected to the laboratory exhaust system.

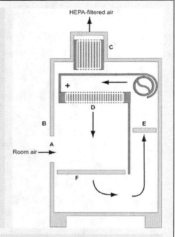

*The tabletop model of a Class II, Type A2 BSC* (A) front opening; (B) sash; (C) exhaust HEPA filter; (D) supply HEPA filter; (E) positive pressure common plenum; (F) negative pressure plenum.. **Note: The A2 BSC should be canopy connected to the exhaust system.**

Figure 30. The Class III BSC.

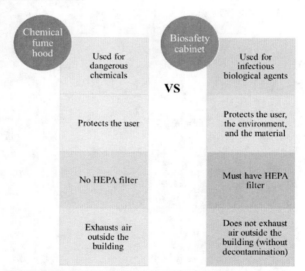

Figure 31. The difference between a fume hood and a biosafety cabinet.

# REFERENCES

[1] ASHRAE Standard 110-2016 *Methods of Testing Performance of Laboratory Fume Hoods*.
[2] Urben, P.G. *Bretherick's Handbook of Reactive Chemical Hazards*: Eighth Edition (2017), pp. 1-1502.

*Chapter 7*

# PERSONAL PROTECTION EQUIPMENT

Personal Protective Equipment is intended for use only when all other measures to eliminate or reduce risks are insufficient or impossible to implement.

A PPE is any device or garment worn by a worker to control the level of risk that can not be controlled or eliminated by providing protection/shield between the hazard and the worker when exposed to:

- Dangerous goods, hazardous chemicals, infectious substances including blood and bodily fluids (BBF);
- Dust, fumes or particles;
- Radiation (ionizing and non-ionizing), ultraviolet or solar radiation;
- Noise;
- Moving objects such as vehicles, trolleys and forklifts;
- Flying objects when using machinery with moving parts;
- Environmental factors, for example, high and low temperature PPE must be used for additional protection when other risk control measures do not provide sufficient exposure control.

## Table 21. PPE selection guide by task

| Task/activity | PPE |
|---|---|
| **Chemicals** | |
| Solids of low or moderate toxicity | Disposable gloves |
| Minimal amounts of liquids (less than 0.1 L) with acute or chronic toxicity | Safety glasses or goggles<br>Appropriate chemical-resistant gloves<br>Clothing covering to knees |
| More than minimal amounts of liquids with acute or chronic toxicity (pure chemicals, mixtures or solutions) | Safety glasses or goggles<br>Chemical-resistant gloves<br>Lab coat<br>Acid-resistant apron if more than 4 liters of highly corrosive chemicals used Consider flame-resistant lab coat if more than 4 liters of flammable liquids used |
| More than minimal amounts of liquids with acute or chronic toxicity (pure chemicals, mixtures or solutions) | Safety glasses or goggles<br>Chemical-resistant gloves<br>Lab coat<br>Acid-resistant apron if more than 4 liters of highly corrosive chemicals used Consider flame-resistant lab coat if more than 4 liters of flammable liquids used |
| Cryogenic liquids | Safety glasses or goggles<br>Face shield required if handling cryovials stored in liquid phase<br>Insulated cryogenic gloves<br>Lab coat recommended |
| Potentially explosive compounds | Safety goggles<br>Face shield<br>Heavyweight gloves<br>Fire-resistant lab coat |
| Pyrophoric (air-reactive) solids or liquids | Safety glasses or goggles<br>Face shield recommended<br>Fire-resistant gloves<br>Appropriate chemical-resistant gloves<br>Fire-resistant lab coat |
| Particularly hazardous substances including carcinogens, reproductive toxins, and reagents of high acute toxicity | Safety glasses or goggles<br>Appropriate chemical-resistant gloves<br>Lab coat<br>Respirators as needed |

*Personal Protection Equipment* 121

| Task/activity | PPE |
|---|---|
| **Biological Materials** | |
| BL1 microorganisms or viruses | Disposable gloves |
| BL2 microorganisms, viruses, viral vectors, human materials or old world primate materials | Disposable gloves<br>Lab coat |
| Procedures outside of the biosafety cabinet without splatter guard when splashes or sprays are anticipated | Safety glasses or goggles<br>Disposable gloves<br>Lab coat |
| **Radiation** | |
| Unsealed radioactive materials or waste | Safety glasses if there is a splash potential<br>Nitrile or other appropriate gloves<br>Lab coat |
| Class 3B or 4 laser | Appropriate eye protection |
| and if UV laser | Gloves<br>Lab coat |
| Laser(s) modified by optics | Appropriate eye protection |
| Open ultraviolet light source<br>and if face enters UV beam<br>and if hand enters UV beam<br>and if body enters UV beam | Safety glasses or goggles with UV protection<br>UV face shield<br>Gloves<br>Lab coat |
| Infrared-emitting equipment | Appropriately-shaded goggles<br>Lab coat |
| **Other Hazards** | |
| Handling hot surfaces and objects such as autoclaved materials and heated glassware | Heat-resistant gloves<br>Lab coat |
| Glassware under pressure or vacuum | Safety glasses or goggles<br>Face shield recommended<br>Lab coat |
| Cutting and connecting glass tubing | Safety glasses or goggles<br>Cut-resistant gloves |
| Sonicator or other loud equipment | Ear plugs |

PPE is one of the least effective methods of controlling occupational risk health and safety risks, according to the control hierachy, and should be used when there are no other practical risk control measures available or identified through a dynamic risk assessment [1]. To perform any operations or experiments, you must decide to wear the appropriate PPE and the various variables, for example:

- The nature of the hazard and the task;
- The compatibility with other PPE;
- The chemical products used, including their concentration and quantity;
- The dangers posed by the chemicals;
- The routes of exposure for the chemical substancies;
- The material of which the PPE is constructed;
- The permeation and degradation rates specific chemicals will have on the material;
- The duration the PPE will be in contact with the chemicals;
- The employer must train the employees before issuing PPE at least in these matters:
  1. When it is necessary to have a PPE.
  2. Which PPE is necessary.
  3. How to wear, doff, adjust and correctly bring the PPE.
  4. Limitations of the PPE.
  5. Adequate care, maintenance, useful life and PPE disposal.

The choice of PPE is always the result of the best possible compromise between the highest level of safety attainable and the need to work in of maximum comfort conditions without being hindered.

## 7.1. HAND PROTECTION

There are several types of gloves that provide protection against and oppose corruption and pervasion to chemicals.

According to the National Ag Safety database, a program supported by NIOSH [2] and the Centers for Disease Control and Prevention, the materials used in the manufacture of gloves designed to provide chemical resistance include the following:

*Personal Protection Equipment*                                    123

- *Butyl* is a synthetic rubber with good resistance to weathering and a wide variety of chemicals.
- *Natural rubber latex* is a highly flexible and conforming material made from a liquid tapped from rubber plants. It is a known allergen.
- *Neoprene* is a synthetic rubber having chemical and wear-resistance properties superior to those of natural rubber.
- *Nitrile* is a copolymer available in a wide range of acrylonitrile content; chemical resistance and stiffness increase with higher acrylonitrile content.
- Polyethylene is a fairly chemical-resistant material used as a freestanding film or a fabric coating.
- *Poly(vinyl alcohol)* is a water-soluble polymer that exhibits exceptional resistance to many organic solvents that rapidly permeate most rubbers.
- *Poly(vinyl chloride)* is a stiff polymer that is made softer and more suitable for protective clothing applications by the addition of plasticizers.
- *Polyurethane* is an abrasion-resistant rubber that is either coated into fabrics or formed into gloves or boots.
- *4H®or Silvershield®* is a registered trademark of North Hand Protection; it is highly chemical-resistant to many different class of chemicals.
- *Viton®*, a registered trademark of DuPont, is a highly chemical-resistant but expensive synthetic elastomer.

When choosing an appropriate glove, consider the required thickness and length of the gloves as well as the material. Confiding in the type and concentration of the chemical, performance characteristics of the gloves, conditions and duration of use, hazards present, and the duration of time a chemical has been in contact with the glove, all gloves must be replaced periodically.

## Table 22. The most common chemical agent-specific glove materials

| Chemical | Glove Material | | | | | |
|---|---|---|---|---|---|---|
| | Butyl | Natural rubber | Neoprene | Nitrile | PVA | PVC |
| Benzene | | | | | | |
| Diesel | | | | | | |
| Gasoline, unleaded | | | | | | |
| Kerosene | | | | | | |
| Hydrochloric Acid (37%) | | | | | | |
| Sulfuric Acid (30-70%) | | | | | | |

*Table adapted from: Forsberg, K. & Mansdorf, S.Z. Quick Selection Guide to Chemical Protective Clothing. 2nd Ed. Van Nostrand Reinhold, NY, NY*

| Not Recommended | Caution (1-4 hours) | Recommended (>4 hours) | Recommended (>8 hours) | Not Tested |
|---|---|---|---|---|

*Chemical substancies can get inside a glove by:*

- Permeation - Diffusion of a chemical through a material on a molecular basis;
- Penetration – Chemical enters through zippers, punctures, or seams;
- Degradation – Chemical causes a change in the physical properties of the material.

# Personal Protection Equipment

## Table 23. The pros and cons of the glove materials that can help lab operator in their proper choice

| Glove Material | Advantages | Disadvantages | Protects Against |
|---|---|---|---|
| Natural latex rubber | Low cost, good physical properties, dexterity, excellent abrasion & tear resistance. | Poor protection against organic solvents, amides, ketones, isocyanates, aldehydes, methylene chloride, oils, greases, brake fluid, or fuels like kerosene & gasoline. Gloves are derived from latex, which can cause allergic contact dermatitis. Poor flame resistance. | Aqueous solutions of acids & bases, Sevin® (carbaryl), organic & inorganic salts & solutions, mercury, ethylene glycol & glycerol. Good for food handling. |
| Polyvinyl chloride (PVC) | Medium cost, very good physical properties, medium chemical resistance. | Plasticizers can be stripped. Poor protection against aldehydes, ketones, isocyanates, hydrocarbons (aliphatic, alicyclic, & aromatic), halogen compounds, heterocyclic compounds, & nitro compounds. | Acids & bases, carbaryl, salts, aqueous solutions, hydrazine, ethylene glycol, glycerol, tricresyl phosphate, polychlorinated biphenyls, mercury, oils, & fats. |
| Neoprene (Polychloroprene) | Medium cost, medium chemical resistance. Excellent protection from physical hazards such as cuts & abrasions. Flexible over a wide temperature range. | Poor protection against isocyanates, hydrocarbons (aliphatic, alicyclic, & aromatic), halogen compounds, & ketones. | Acids, bases, carbaryl, alcohols, mercury, polychlorinated biphenyls, hydrazine, chlorine gas, oils, greases, petrochemicals, & some aldehydes. |
| Nitrile | Low cost, excellent physical properties & dexterity. | Poor protection against methylene chloride, trichloroethylene, aromatic hydrocarbons, ketones, & acetates. | Oils, greases, hydraulic oil, motor oil, weak acids & bases, hydrazine, carbaryl, aliphatic & alicyclic hydrocarbons, fuels, pesticides, & mercury. |
| Butyl | Synthetic rubber with good resistance to a wide variety of chemicals. Highly resistant to gases. | Expensive. Poor protection against aliphatic & aromatic hydrocarbons (including gasoline), & chlorinated solvents. | Glycol ethers, aniline, ketones, some esters, acids, bases, acetates, alcohols, acetone, some amines, hydrazine, hydroxyl compounds, isocyanates, tricresyl phosphate, mercury, peroxides, polychlorinated biphenyls, nitro compounds, chlorine gas, & aldehydes. |
| Polyvinyl alcohol (PVA) | Durable. Resists a very broad range of organic compounds. Flexible over a wide temperature range. | PVA is water soluble - do not use in water or aqueous solutions. Poor protection against alcohols. | Aliphatic & aromatic hydrocarbons, methylene chloride, isocyanates, perchloroethylene, trichloroethylene, some ketones (not acetone or methyl ethyl ketone), tricresyl phosphate, xylene, & some esters. |
| Fluoroelastomer (Viton) ™ | Specialty glove. Highly resistant to gases. | Expensive. Poor physical properties - minimal resistance to cuts & abrasions. Poor protection against most ketones. | Acids, bases, chlorinated hydrocarbons, aromatic & aliphatic hydrocarbons, hydroxyl compounds, glycol ethers, some esters, tricresyl phosphate, trichloroethylene, aniline, perchloroethylene, mercury, styrene, toluene, alcohols, polychlorinated biphenyls, & some amines. |

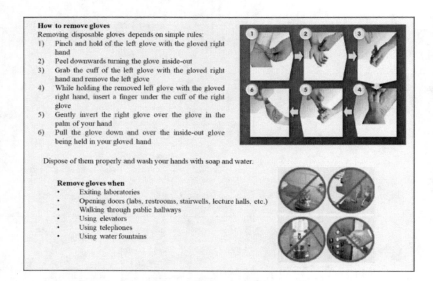

Figure 32. Illustration of the sequence of operations for the correct removal of laboratory gloves.

*Glove Care*
- Inspect gloves before use for tears, excessive wear, and punctures;
- Store in a clean, dry location;
- Discard leather and cloth gloves if they become saturated with oil or other chemicals;
- Leak test chemical gloves by sealing the wrist and filling the glove with air -Use a clean plastic tube or low pressure air line – not your mouth!

*Hand Care*
- Avoid washing your hands with solvents, harsh soaps, or abrasives;
- Clean and bandage all cuts and abrasions;
- Immediately remove any imbedded foreign materials;
- Wash immediately after using any chemical – Even if you did not detect leakage;
- Pay attention to skin rashes—get an immediate medical evaluation;
- Wear cotton gloves under rubber gloves to reduce sweating.

## 7.2. Eye and Face Protection

Employees can be exposed to a large number of hazards that pose danger to their eyes and face. OSHA requires employers to ensure that employees have appropriate eye or face protection if they are exposed to eye or face hazards from [3]:

- Flying particles;
- Molten metal;
- Liquid chemicals, acids or caustic liquids;
- Chemical gases or vapors;
- Potentially infected material or potentially harmful light radiation.

Eye protection should be worn at all times while working with hazardous chemicals/biological materials or any physical hazards in the laboratory. Visitors should be provided with temporary protective goggles or, at least, protective glasses if they are allowed in any area where the occupational use of eye protection is required. Employees with prescription lenses must:

- Incorporate the prescription in eye protection or;
- Wear eye protection over prescription lenses;
- Without disturbing the correct position of prescription lenses or the protective lenses.

The use of contact lenses is not recommended while working with chemicals that cause eye irritation. In case of chemical accident in the eyes, there might be some protection but, on the other hand, the presence of the lens would be an impediment to rapid and accurate flushing of the eyes. The lens should be removed which could cause in damage to the eye itself.

If, however, the user of contact lens conscientiously wears a pair of good-quality goggles at any time when there is the possibility of an accident occuring, there is probably little risk in wearing the contact lens.

128 *Maria Pia Gatto*

Even in the latter case, where extremely corrosive vapors are likely to be involved, there is a possibility of capillary action causing these vapors to be drawn under the contact lens, and the wearer should be careful if there is any suspicion that this could happen.

### Safety Glasses

Safety glasses provide eye protection from moderate effect and particles related with grinding, sawing, scaling, broken glass, and minor chemical splashes, and so on. Side defenders are required when there is a risk of flying objects.

In the form of prescription for those people requiring corrective lenses, safety glasses are accessible. In the case of safety glasses do not give sufficient insurance for procedures that include substantial synthetic utilize, such as, mixing, pouring, or blending, splash goggles should be used.

### Splash and Laser Goggles

Including the potential dangers of chemical splashes, the use of concentrated corrosive material and the bulk chemical transfer spray goggles give satisfactory ocular protection from numerous hazards. The goggles are available with transparent or colored lenses, fog-proof and vented or unvented frames. Be aware that goggles intended for carpentry (can be recognized by the various small holes throughout the face piece) are not suitable for working with chemicals. In case of a splash, the chemicals could enter into these small holes, and result in a chemical exposure to the face. Make sure that goggles you choose are evaluated for use with chemicals.

The lens of the eyewear is a filter/absorber designed to reduce the transmittance of light to a specific wavelength. The lens can filter (or absorb) a specific wavelength while maintaining adequate light transmission for other wavelengths. A single pair of safety glasses is not available for protection from all LASER outputs. The type of eye protection required is dependent on the spectral frequency or specific wavelength of the laser source.

Make sure that the glasses you chooseare evaluated for use with chemicals. The lens of the eyewear is a filter/absorber designed toreduce the

*Personal Protection Equipment* 129

transmittance of light to a specific wavelength.The lens can filter (or absorb) a specific wavelength whilemaintaining adequate light transmission for otherwavelengths. A single pair of safety glasses is not availablefor protection from all LASER outputs. The type of eyeprotection required depends on the spectral frequency orthe specific wavelength of the laser source.

## *Face Shields*

When utilized in combination with safety glasses or splash goggles, face shields provide additional protection to the eyes and face. Face shields comprise of a flexible headgear and face shield of tinted or clear lenses or a mesh wire screen. When the whole face needs assurance, they ought to be utilized as a part of operations and worn to shield the eyes and face from flying particles, metal sparks, and chemical/biological splashes. Face shields with a mesh wire screen are not appropriate for use with chemicals. Face shields must not be used alone and are not a substitute for appropriate eyewear and they should always be worn in conjunction with an essential type of eye protection, for example, safety glasses or splash goggles.

# 7.3. RESPIRATORY PROTECTION

As pointed out at the beginning of this chapter, the primary objective is to prevent atmospheric contamination respiratory hazards: dusts, mists, fogs, fumes, sprays, smokes or vapors. The engineering controls are the 1st priority means of control the exposure to hazards by:

- Enclosure or confinement of the operation;
- General and local ventilation, and
- Replacement with less toxic materials.

Only if the engineering controls are not achievable should be used respirators. If correctly selected and used, respirators can protect the user from:

130    *Maria Pia Gatto*

- Fumes and smokes (welding vapours);
- Harmful dusts (lead, silica, and other heavy metals);
- Gases and vapors (chemical exposures);
- Lack of oxygen (oxidation, displacement, and consumption);
- Biological hazards (tuberculosis, whooping cough, flu viruses).

The different types of masks are listed below:

## Dust Mask

The use of the term "dust" mask for the non-rigid soft felt mask is somewhat of a misnomer since, in modified forms, they can be used for other applications such as limited protection against paint fumes, moderate levels of organics, acid fumes, mercury, etc., although their greatest use is against dust nuisance.

For individual filter respirators it is necessary to have the following markings:

1. Name of approval holder/manufacturer business name, a registered trademark, or an easily understood abbreviation of the applicant/approval holder's business name as recognized by NIOSH. When applicable, the name of the entity to which the FFR has been private labeled by the approval holder may replace the approval holder business name, registered trademark, or abbreviation of the approval holder business name as recognized by NIOSH.
2. NIOSH in block letters or the NIOSH logo.
3. NIOSH Testing and Certification approval number, e.g., TC-84A-XXXX.
4. NIOSH filter series and filter efficiency level, e.g., N95, N99, N100, R95, P95, P99, P100.

## Personal Protection Equipment

5. Model number or part number: The approval holder's respirator model number or part number, represented by a series of numbers or alphanumeric markings, e.g., 8577 or 8577A.

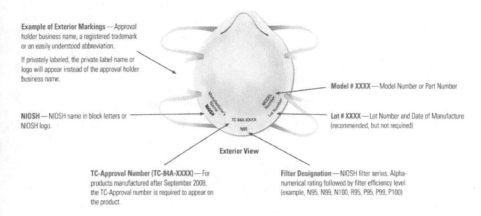

Figure 33. Required Labeling of NIOSH-Approved N95 Filtering Facepiece Respirators (2014).

## Half-Face Respirator

### Table 24. Chemical cartridge color coding

| Color | Type |
|---|---|
| White | Acid Gas |
| Black | Organic Vapors |
| Green | Ammonia Gas |
| Yellow | Acid Gas & Organic Vapor |
| Olive | Mulit-gas (protects against numerous gases and vapors) |
| Magenta | Particulate Filter Cartridge (HEPA) (Also called P100) |

The half-face cartridge respirator is the type most frequently used, especially in atmospheres in which there is little or no problem of irritation or absorption of material through the skin. All manufacturers use the same color coding for gas/vapor protection.

## Full-Face Air-Purifying Respirator

Full-face air-purifying respirators are similar in many respects to half-face respirators, with the obvious difference that the mask covers the upper part of the face, protecting the eyes.

Users must inspect their respirators before and after use. Respirator inspections must include control:

- the sealing surface is clean and free of cracks and holes;
- the rubber and the elastic parts have good malleability and no signs of deterioration;
- the inhalation and exhalation valves are clean and well seated;
- the straps are sufficiently elastic and free of areas worn;
- if full face, face shield is clean and transparent (no smudges, scratches, or other damage that may impede visibility);
- those respirators that fail inspection must be removed from service and replaced.

Before using a respirator, the user must perform a positive and negative pressure check. The wearer must ensure that the present face conditions allow an effective sealing (for example the user must be clean shaved).

*Positive pressure check.* Close off the exhalation valve with palms and exhale gently. No leakage outward around the seal should occur.

*Negative pressure check.* Close off the cartridges and inhale. The respirator should collapse slightly on the face. No leakage around the face seal should occur while maintaining a negative pressure inside the respirator for several seconds.

Respirators must be cleaned and disinfected after each use (Figure 34). These procedures are provided for employer use when cleaning the respirators. They are of a general nature and must ensure that the respirator is properly cleaned and disinfected in such a way as to avoid damage to the respirator and does not cause injury to the user.

## Personal Protection Equipment

**Procedures for Cleaning Respirators**

A. Remove filters, cartridges, or canisters. Disassemble facepieces by removing speaking diaphragms, demand and pressure- demand valve assemblies, hoses, or any components recommended by the manufacturer. Discard or repair any defective parts.

B. Wash components in warm (43 deg. C [110 deg. F] maximum) water with a mild detergent or with a cleaner recommended by the manufacturer. A stiff bristle (not wire) brush may be used to facilitate the removal of dirt.

C. Rinse components thoroughly in clean, warm (43 deg. C [110 deg. F] maximum), preferably running water. Drain.

D. When the cleaner used does not contain a disinfecting agent, respirator components should be immersed for two minutes in one of the following:

1. Hypochlorite solution (50 ppm of chlorine) made by adding approximately one milliliter of laundry bleach to one liter of water at 43 deg. C (110 deg. F); or,

2. Aqueous solution of iodine (50 ppm iodine) made by adding approximately 0.8 milliliters of tincture of iodine (6-8 grams ammonium and/or potassium iodide/100 cc of 45% alcohol) to one liter of water at 43 deg. C (110 deg. F); or,

3. Other commercially available cleansers of equivalent disinfectant quality when used as directed, if their use is recommended or approved by the respirator manufacturer.

E. Rinse components thoroughly in clean, warm (43 deg. C [110 deg. F] maximum), preferably running water. Drain. The importance of thorough rinsing cannot be overemphasized. Detergents or disinfectants that dry on facepieces may result in dermatitis. In addition, some disinfectants may cause deterioration of rubber or corrosion of metal parts if not completely removed.

F. Components should be hand-dried with a clean lint-free cloth or air-dried.

G. Reassemble facepiece, replacing filters, cartridges, and canisters where necessary.

H. Test the respirator to ensure that all components work properly.

Figure 34. Best management practices for cleaning respirators.

## 7.4. CHEMICAL PROTECTIVE CLOTHING

Chemical protective clothing (CPC), which includes gloves, aprons, coveralls, pants, jackets, and boots, is a subset of PPE and consist of all items of protective clothing whose main purpose is to provide skin protection against chemical, physical, and/or biological hazards. Many stressors pose "invisible" hazards and offer no warning properties. Unfortunately, no single combination of protective equipment and clothing can protect against all hazards.

Therefore, CPC should be used with other protective methods, such as engineering controls, to restrict exposure. The use of CPC can create dangers to the wearer, such as heat stress, physical and psychological stress, impaired vision and restricted mobility and communication. In general, the higher the CPC level, the higher the risk associated. For each situation, CPC should be selected which provides an adequate level of security.

## Table 25. Chemical-protective clothing types

| Chemical hazard | | Type 1 | Gas-thight protection against chemicals and vapors and toxic particles |
|---|---|---|---|
| | | Type 2 | Non gas-thight protection |
| | | Type 3 | Protection against pressurized liquid chemicals |
| | | Type 4 | Protection against liquid aerosols |
| | | Type 5 | Protection against airborne solid particulate chemicals |
| | | Type 6 | Limited protection against liquid mist |

Over-protection may be hazardous and should be avoided. The CPC should be worn whenever there are potential hazards arising from direct exposure. Examples include: emergency response; loss of equipment or faults; chemical treatment processes, such as chemical baths; clean and disposal of the hazardous waste site; operations that produce particulate hazards, including asbestos removal; and the application of pesticides.

The chemical-protective clothing is subdivided into 6 types: to obtain the certification corresponding to one of these types, the garments must pass tests of movement and sealing inside a cabin in which they are kept in

*Personal Protection Equipment* 135

contact with liquids, particles and gases [4]. The level of exposure determines the choice of the type of garment.

Protective clothing can be classified according to design, performance, and service life.

- *Design.* The categorization of clothing by design is mainly a means of describing which areas of the body the clothing item is intended to protect. In the emergency response, the cleaning of hazardous waste site and hazardous chemical operations, the only acceptable types of protective clothing include fully or totally encapsulating suits and nonencapsulating or "splash" suits, as well as accessory clothing items, such as chemically resistant gloves or boots. These descriptions apply to how the clothing is designed and not to its performance.
- *Performance.* The National Fire Protection Association (NFPA) has classified suits by their performance as:
  1. Vapor-protective suits (NFPA Standard 1991) provide "gas-tight" integrity and are intended for response situations where no chemical contact is permissible. This type of suit would be equivalent to the clothing required in EPA's Level A.
  2. Liquid splash-protective suits (NFPA Standard 1992) offer protection against liquid chemicals in the form of splashes, but not against continuous liquid contact or chemical vapors or gases. Essentially, the type of clothing would meet the EPA Level B needs. It is important to note, however, that by wearing liquid splash-protective clothing, the wearer accepts exposure to chemical vapors or gases because this clothing does not offer gas-tight performance. The use of duct tape to seal clothing interfaces does not provide the type of wearer encapsulation necessary for protection against vapors or gases.
  3. Support function protective garments (NFPA Standard 1993) must also provide liquid splash protection but offer limited physical protection. These garments may comprise several separate protective clothing components (i.e., coveralls, hoods,

gloves, and boots). They are intended for use in non-emergency, non-flammable situations where chemical hazards have been fully characterized. Examples of support functions include proximity to chemical processes, decontamination, hazardous waste clean up, and training. Protective clothing for the support function must not be used in chemical emergency response or in situations where chemical hazards remain uncharacterized.

4. These NFPA standards define minimum performance requirements for the manufacture of chemical protective suits. Each standard requires rigorous testing of the suit and the materials that comprise the suit in terms of overall protection, chemical resistance, and physical properties. The manufacturer as meeting the requirements of the respective NFPA standard may label suits that are compliant with an independent certification and testing organization. Manufacturers must also provide the documentation that show all the test results and the characteristics of their protective clothing.

5. Protective clothing should completely cover both the user and his or her breathing apparatus. In general, respiratory protective equipment are not designed to resist chemical contamination.

6. Level A protection (vapor-protective suits) require this configuration. Level B ensembles may be configured either with the SCBA on the outside or inside. However, it is strongly recommended that the wearer's respiratory equipment be worn inside the ensemble to prevent its failure and to reduce decontamination problems. Level C ensembles use cartridge or canister type respirators which are generally worn outside the clothing.

The CPC' ability to act as a chemical barrier is determined by the CPC material and the method of construction [5]. Usually, each chemical interacts with a given plastic or elastomer differently so there is a unique situation for each chemical/CPC material pair [6]. Ideally, the chosen material(s) must be based on:

## Personal Protection Equipment

- *Permeation.* Process by which a chemical moves through a material on a molecular basis. The Permeation rate is usually expressed in terms of amount of chemical, which passes through a given area per unit time (milligrams per square meter per minute). The total amount of chemical permeating a CPC material is dependent on the area exposed and the duration of exposure. For a given chemical/CPC material pair, the permeation rate decreases as the material thickness increases.
- *Breakthrough time.* The elapsed time from the initial contact of the chemical with the outside surface of the CPC material to the first detection of the chemical on the inside surface of the material. There may be situations in which breakthrough times are longer in a chemical/ CPC pair than another, however the material with a shorter breakthrough time is recommended because its permeation rate is very small compared to pair of material chemical/ CPC with the long breakthrough time. On the other hand, the breakthrough time may be the most important criterion when the chemical is carcinogenic and no skin contact is desired. Most breakthrough time and permeation rate data has been determined by using the American Society for Testing and Materials (ASTM) F739 Standard (reference 10-9).
- *Degradation.* Degradation is physical changes in a material as the result of a chemical/ physical exposure or use. The most common observation of material degradation is discoloration, swelling, loss of physical strength, or deterioration.
- *Penetration.* The gross movement of a chemical through zippers, seams, or imperfections in CPC material.
- *Chemical mixtures.* Mixtures of chemicals can be significantly more aggressive on materials than any single chemical alone. One chemical may pull another with it through the material. Another may change the CPC material structure and allow greater diffusion of other chemicals. Reference 10-10 reports protective clothing chemical resistance data, listing over 10,500 chemical permeation tests on chemical protective clothing exposed to 860 chemicals and

mixtures of chemicals. It also contains over 3,000 chemical degradation tests.

> There are no design or test criteria specified in regulations or guidelines specific to lab coats, but generally:
> - ✓ Lab coats are not tested for normal conditions that may be experienced in a research lab with respect to chemical use, or joined research activities.
> - ✓ Manufacturers of the lab coats do not provide information about the capability of a lab coat for a combination of hazards. If a coat is "flame resistant", it may not be chemical resistant or acid resistant.
> - ✓ If a coat is sold as flame resistant, this means it is not tested involving flammable chemicals on the coat. The flame resistance test criteria includes simulation of the possibilities of a flash
> - ✓ Lab coats should be provided for protection and convenience. They should be worn at all times in the lab areas.
> - ✓ Due to the possible absorption and accumulation of chemicals in the material, lab coats should not be worn in the lunchroom or elsewhere outside the laboratory.

Figure 35. Some useful tips on the correct use of the lab coat.

## 7.5. POTENTIAL LIMITS OF PPE

The limits of PPE may be summarised in the following [7]:

*For the Individual*
- Reduced sensory input:
  - field of view restricted by the visor of air purifiers respirators;
  - reduced sense of tact because of the thichness of the glove;
  - reduction of hearing from noise moving in suits with chemically-esistant fabric and by blowing units
- greater communication difficulties. This is twice– muffled speech and hearing difficulty. This combined with the reduced sensory input reduces the capacity to maintain situational awareness.
- increased risk of tripping due to unfamiliar footwear and surface conditions, as well as water under the feet.
- Psychological distress:
  - due to claustrophobia;

## Personal Protection Equipment

139

- – due to lack of confidence in the protective performance of the PPE;
- – appearing more confronting to patients; Because of stress of operating in a "less familiar way".
- Potential heat related illness due to the relatively impervious fabric and complete coverage of most ensembles preventing evaporation of perspiration. This may manifest as slurred speech, staggering gait, or altered behaviour.
- Difficulty in fitting and exclusion of those with glasses or facial hair from some configurations of PPE.
- Risk of latex allergy with some materials in respirator options.
- The impact of work performance as a sum of the impacts of previous points. This limits the types of care that can be delivered and reduces the time spent working in PPE.
- Adequacy of the procedures for wearing and levation.
- A security supervisor can oversee the management of operations in PPE and monitor personnel duration spent in PPE. A 'buddy' system, with staff working in pairs at all times when in PPE, will help in monitor staff more closely for the first signs of heat illness. Training increases familiarity with the PPE and allows you to address some of these limitations. Training is mandatory to ensure safe and effective operations in PPE.

*Organisational*
- Initial cost of equipment and recurring expenses as the shelf-life of the components is reached.
- Maintenance and periodic inspection of equipment.
- Storage of the equipment:
  - – To preserve the actual integrity of the performance, the PPE must be stored in a temperature controlled environment, away from moisture and in such a way as to minimise damage;
  - – Sufficient equipment for the management of mass casualties requires a significant amount of space.

- Training burden:
  - Have sufficient personnel trained to provide 24/h coverage;
  - Maintaining competence with a requirement to retrain at least once a year;
  - Standardised training for consistency and interoperability.
- Develope and promote standards for safe funtioning of PPE.

## REFERENCES

[1] Gudgin Dickson, E.F. (2012) *Personal Protective Equipment for Chemical, Biological, and Radiological Hazards: Design, Evaluation, and Selection*, pp. 1-331. DOI: 10.1002/9781118422991.

[2] National Institute for Occupational Safety and Health (NIOSH): *Indexed Dermal Bibliography* (1995-2007), NIOSH Publication No. 2009–153: Cincinnati, Ohio: NIOSH, 2009.

[3] Occupational Safety and Health Administration (OSHA) *Personal Protective Equipment*. OSHA 3151-12R (2004).

[4] Reese, C.D., Eidson, J.E. (2006) *Handbook of OSHA Construction Safety and Health* Boca Raton CRC Press eBook ISBN9781420006230.

[5] Forsberg, K. and Keith, L. H. (1999) *Chemical Protective Clothing: Performance Index*, 2nd ed. Cincinnati, Ohio: ACGIH®.

[6] *Lab Coat Selection, Use, and Care at MIT*. (2013, September). Retrieved from https://labcoats.mit.edu/guidance

[7] Occupational Safety and Health Administration (OSHA). *Major Requirements of Osha's Respiratory Protection Standard 29 CFR*, 1910.134 OSHA Office of Training and Education Rev. December 2006.

*Chapter 8*

# LABORATORY FIRE SAFETY

Laboratories, especially those using solvents in any quantity, have the potential for flash fires, explosion, rapid spread of fire, and high toxicity of products of combustion (heat, smoke, and flame). Fire is the most common serious danger you face in a typical laboratory. While proper procedures and training can minimize the chances of an accidental fire, laboratory workers should still be prepared to deal with a fire emergency should it occur. When dealing with a laboratory fire, all containers of infectious materials must be placed in autoclaves, incubators, refrigerators, or freezers for containment [1].

## 8.1. CLASSIFICATION OF FIRE, FLAMMABLE MATERIALS AND EXTINCTION SYSTEMS

Figure 36 shows the triangle of fire that actually, is a tetrahedron, because there are four elements that need to be present for a fire to exist. There must be oxygen to sustain combustion, heat to raise the material to its ignition temperature, fuel to support the combustion and a chemical reaction among the other three elements. Remove any one of the four elements to extinguish the fire. *The concept of fire protection is based upon keeping these four elements separate.*

Figure 36. The fire triangle.

**Table 26. A, B, C, D, and K fire classes, symbols and fuel types**

| Fire Class | Symbol | Fuel type |
|---|---|---|
| A | | Class A fires are fires in ordinary combustibles such as wood, paper, cloth, rubber, and many plastics. |
| B | | Class B fires are fires in flammable liquids such as gasoline, petroleum greases, tars, oils, oil-based paints, solvents, alcohols. Class B fires also include flammable gases (i.e., propane and butane) but do not include fires involving cooking oils and grease. |
| C | | Class C fires are fires involving energized electical equipment such as computers, servers, motors, transformers, and appliances. Remove the power and the Class C fire becomes one of the other classes of fire. |
| D | | Class D fires are fires in combustible metals such as magnesium, titanium, zirconium, sodium, lithium, and potassium. |
| K | | Class K fires are fires in cooking oils and greases such as animal and vegetable fats. |

*Laboratory Fire Safety* 143

Not all fires are the same; combustion may be classified in one or more of the following fire classes and the right fire extinguisher size and agent for the hazard must be selected [2].

## 8.2. FIRE EXTINGUISHERS

Any type of fire, whether originating from ordinary combustibles, flammable liquids or electric sources, has a specific counteracting agent that allows it to be extinguished. Therefore, there are many special classes of fire extinguishers with a unique extinguishing agent tailored specifically for each blaze. This allows fires to extinguish quicker and more easily than if there was only one fire-suitable-all extinguishing agent. The National Fire Protection Association (NFPA) highlights the standard for the selection, use and maintenance of a fire extinguisher. Apart from the type of fires they dissolve, the fire extinguishers are also numerically classified by the size of the fire they can control. Standard classifications act as quick identifiers for the scale of fire which can be handled.

To start with, numerical ratings only apply to classes "A" and "B". A rating like this would appear on the fire extinguisher label –"2-A:10-B:C". Here, the numerical rating precedes the class letter, in this case, "2". When the numeral rating is multiplied by 1.25, it's the equivalent capacity in gallons of water. he letter "C" is not accompanied by any rating, as its sole purpose is to show that the extinguishing agent does not conduct electricity. Fire-extinguishing capacity is rated according to ANSI/UL 711, Rating and Fire Testing of Fire Extinguishers. Other details about the extinguisher such as class category, are mentioned on fire extinguisher labels—without which, determining the right fire extinguisher would be impossible. In addition to labels, proper fire extinguisher signs should be erected, and in fact, are a safety requirement by OSHA.

## Table 27. Fire extinguishers description

| Types of Fire Extinguishers | Description | Use |
|---|---|---|
| Water and Foam | Water and Foam fire extinguishers extinguish the fire by taking away the heat element of the fire triangle. Foam agents also separate the oxygen element from the other elements. | Water extinguishers are for Class A fires only - they should not be used on Class B or C fires. The discharge stream could spread the flammable liquid in a Class B fire or could create a shock hazard on a Class C fire. |
| Carbon Dioxide | Carbon Dioxide fire extinguishers extinguish fire by taking away the oxygen element of the fire triangle and be removing the heat with a very cold discharge. | Carbon dioxide can be used on Class B & C fires. They are usually ineffective on Class A fires. |
| Dry Chemical | Dry Chemical fire extinguishers extinguish the fire primarily by interrupting the chemical reaction of the fire triangle. | Today's most widely used type of fire extinguisher is the multipurpose dry chemical that is effective on Class A, B, and C fires. This agent also works by creating a barrier between the oxygen element and the fuelclement on Class A fires. Ordinary dry chemical is for Class B & C fires only. It is important to use the correct extinguisher for the type of fuel! Using the incorrect agent can allow the fire to re-ignite after apparently being extinguished succesfully. |
| Wet Chemical | Wet Chemical is a new agent that extinguishes the fire by removing the heat of the fire triangle and prevents re-ignition by creating a barrier between the oxygen and fuel elements. | Wet chemical of Class K extinguishers were developed for modern, high efficiency deep fat fryers in commercial cooking operations. Some may also be used on Class A fires in commercial kitchens. |
| Halogenated Agent | Halogenated or Clean Agent extinguishers include the halon agents as well as the newer and less ozone depleting halocarbon agents. They extinguish the fire by interrupting the chemical reaction and/or removing heat from the fire triangle. | Halogenated Agent extinguishers are effective on Class A, B and C fires. Smaller sized handheld extinguishers are not large enough to obtain a 1A rating and may carry only a Class B and C rating. |

| Types of Fire Extinguishers | Description | Use |
| --- | --- | --- |
| Dry Powder | Dry Powder extinguishers are similar to dry chemical except that they extinguish the fire by separating the fuel from the oxygen element or by removing the heat element of the fire triangle. | However, dry powder extinguishers are for Class D or combustible metal fires, only. They are ineffective on all other classes of fires. |
| Water Mist | Water Mist extinguishers are a recent development that extinguish the fire by taking away the heat element of the fire triangle. They are an alternative to the clean agent extinguishers where contamination is a concern. | Water mist extinguishers are primarily for Class A fires, although they are safe for use on Class C fires as well. |
| Cartridge Operated Dry Chemical | Cartridge Operated Dry Chemical fire extinguishers extinguish the fire primarily by interrupting the chemical reaction of the fire triangle. | Like the stored pressure dry chemical extinguishers, the multipurpose dry chemical is effective on Class A, B, and C fires. This agent also works by creating a barrier between the oxygen element and the fuel element on Class A fires. Ordinary dry chemical is for Class B & C fires only. It is important to use the correct extinguisher for the type of fuel! Using the incorrect agent can allow the fire to re-ignite after apparently being extinguished successfully. |

## 8.3. Rules for Employers in Case of Lab Fires

In general, small bench-top fires in laboratory spaces are not uncommon. Large laboratory fires are rare. However, the risk of severe injury or death is significant because fuel load and hazard levels in labs are typically very high.

According to OSHA regulations lab employers should:

1. *Ensure that workers are trained to do the following in order to prevent fires.*
   - Plan work. Have a written emergency plan for your space and/or operation.
   - Minimize materials. Have present in the immediate work area and use only the minimum quantities necessary for work in progress. Not only does this minimize fire risk, it reduces costs and waste.
   - Observe proper housekeeping. Keep work areas uncluttered, and clean frequently. Store unnecessary materials promptly. Keep aisles, doors, and access to emergency equipment unobstructed at all times.
   - Observe restrictions on equipment (i.e., keeping solvents only in an explosion-proof refrigerator).
   - Keep barriers in place (shields, hood doors, lab doors).
   - Wear proper clothing and personal protective equipment.
   - Avoid working alone.
   - Store solvents properly in approved flammable liquid storage cabinets.
   - Shut door behind you when evacuating.
   - Limit open flames use to under fume hoods and only when constantly attended.
   - Keep combustibles away from open flames.
   - Do not heat solvents using hot plates.
   - Remember the "RACE" rule in case of a fire.

*Laboratory Fire Safety* 147

R = Rescue/remove all occupants
A = Activate the alarm system
C = Confine the fire by closing doors
E = Evacuate/Extinguish

2. *Ensure that workers are trained in the following emergency procedures:*
   - Know what to do. You tend to do under stress what you have practiced or pre-planned. Therefore, planning, practice and drills are essential.
   - Know where you will find the nearest fire extinguisher, fire alarm box, exit(s), telephone, emergency shower/eyewash, and first-aid kit, etc.
   - Be aware that emergencies are rarely "clean" and will often involve more than one type of problem. For example, an explosion may generate medical, fire, and contamination emergencies simultaneously.
   - Train workers and exercise the emergency plan.
   - Learn to use the emergency equipment provided.

3. *Train workers to remember the "PASS" rule for fire extinguishers:*
   - PASS summarizes the operation of a fire extinguisher.
   - P – Pull the pin
     A – Aim extinguisher nozzle at the base of the fire
     S – Squeeze the trigger while holding the extinguisher upright
     S – Sweep the extinguisher from side to side; cover the fire with the spray

4. *Train workers on appropriate procedures in the event of a clothing fire:*
   - If the floor is not on fire, STOP, DROP and ROLL to extinguish the flames or use a fire blanket or a safety shower if not contraindicated (i.e., there are no chemicals or electricity involved).

- If a coworker's clothing catches fire and he/she runs down the hallway in panic, tackle him/her and smother the flames as quickly as possible, using appropriate means that are available (e.g., fire blanket, fire extinguisher).

## 8.4. EVACUATION PROCEDURE IN CASE OF FIRE

Each laboratory must prepare an emergency plan and all personnel should be familiar with it [3]. This emergency plan should include:

1. An inventory that includes the quantities and locations of all flammable, pyrophoric, oxidizing, toxic, corrosive, reactive, radioactive materials, nonionizing radiation, biological materials, and compressed and liquefied gases.
2. A list of responsible personnel who are designated and trained to be liaison personnel for the fire department or other emergency responders.
3. Action to be taken by laboratory personnel upon activation of the fire alarm. This should include instructions to turn off flames and other ignition sources, close the fume hood sash, close all hazardous materials containers, and turn off all electrical equipment. All staff are required to exit the building when the fire alarm is activated.
4. Location of emergency equipment in the laboratory (fire extinguishers, emergency shower, eyewash, spill kit and fire blanket if available).
5. Procedures for extinguishing clothing fires (stop, drop & roll, cover face with hands and use fire blanket, do not use fire extinguisher), using emergency shower and eyewash and spill kits.
6. Primary and secondary evacuation routes to the outside of the building.
7. Identify an area outside of the building to meet and account for all laboratory personnel.

# Laboratory Fire Safety 149

8.  Instructions not to reenter the building until qualified Emergency Responders provide notification that it is safe to return.

## 8.5. LABORATORY FIRE SAFETY COMPLIANCE CHECKLIST

### Table 28. Fire safety checklist

| | General fire safety | Yes | No | N/A |
|---|---|---|---|---|
| 1 | Exit signs are lit and emergency lights operational. *Emergency signs help direct individual out of a building and emergency lighting provides minimal lighting levels in case of a power failure. Report any fixture that is not working to your Building Manager.* | | | |
| 2 | Staff knows where multi-purpose, $CO_2$ or pressurized water fire extinguisher present and fully charged (within 50 feet of any point). *Labs that use substantial quantities of flammable hazardous chemicals will have a multi-purpose (ABC) fire extinguisher mounted inside the laboratory. Additional fire extinguishers will also be found in the hallways. If the fire extinguisher is used or found to be not fully charged, immediately contact EH&S for replacement.* | | | |
| 3 | New or surplus equipment, trash, and empty containers not discarded in the corridor. *Corridors are intended to provide a safe and efficient means of exiting a building in emergencies and during normal daily activities. They should not be used as a storage area at any time.* | | | |
| 4 | Laboratory doors remain closed at all time. *Building ventilation systems and fume hood designs depend on laboratory doors to remain closed at all time. Doors left open can render a fume hood useless, exposing building occupants to hazardous chemicals.* | | | |
| 5 | Warning signs are listed on the door of the lab (ex. Flammable solvents, biohazard, etc.). *Beware of any unusual chemical, biological or physical hazard needed to be prominently posted in or near all laboratory entrance doors.* | | | |
| 6 | Emergency evacuation routes and outside meeting point are posted. *Evacuation routes from each laboratory to the two closest exits and the area where everyone is to meet for a "head-count" must be posted.* | | | |
| 7 | Emergency procedures are written and available. *Alarm activation, evacuation and building re-entry procedures, clothing fires, and equipment shutdown procedures or applicable emergency operation must be written and readily available to all laboratory occupants.* | | | |

## Table 28. (Continued)

| General fire safety | Yes | No | N/A |
|---|---|---|---|
| 8 | Equipment maintenance plans are written. *Maintenance plans for all equipment used in a laboratory must be written and available.* | | | |
| 9 | Aisles free of clutter (no tripping hazards) and exit doors not blocked. *Generally, all aisles leading to fire exits must be at least 36 inches wide in laboratories. Equipment and furniture must be placed to prevent any obstruction to the fire exits. Any space over 1,000 square feet must have two fire exits.* | | | |
| 10 | A current inventory and MSDSs of all chemicals used is available. *All hazardous materials must be listed on an inventory associated with the MSDS collection. The chemical supplier, manufacturer, or distributor should accompany the chemical name. DOT hazard class and NFPA ratings for all hazardous chemicals should also be included in the inventory.* | | | |
| 11 | Laboratory fume hoods have current inspection labels. *All fume hoods must be inspected annually by EH&S and have a current inspection sticker posted on the facing of the hood.* | | | |
| 12 | Quantity of flammable/combustible liquids does not exceed storage limits. *No more than 25 gal. of flammable liquids can be stored in a lab outside of a flammable storage cabinet. Maximum allowable containers for flammable/combustible liquids: 4 L (1.1 gal) glass or 20 L (5 gal) metal.* | | | |
| 13 | Refrigerators for flammable are explosion proof type and are properly marked. *Residential type refrigerators cannot be used to store flammable liquids. The refrigerator must be rated "laboratory safe".* | | | |
| **Gas Cylinders** | | | |
| 14 | Number of compressed gas cylinders does not exceed the maximum number allowed. *Maximum number of compressed gas cylinders per lab: No more than 6 flammable or oxidizing gases; No more than 3 flammable gases and no more than 3 gases with an NFPA Health Hazard Rating of 3.* | | | |
| 15 | All cylinders not in use are stored in an appropriate location. *Cylinders, empty or full, may not be stored in a corridor.* | | | |
| 16 | All cylinders are properly secured. *Gas cylinders must be anchored by chains, clamps, or stands.* | | | |
| 17 | All cylinders without regulators are capped. *Cylinders not in current use must have the regulator removed and the cap secured.* | | | |

## Laboratory Fire Safety

|  |  | Yes | No | N/A |
|---|---|---|---|---|
| | **Chemical Storage** | | | |
| 18 | Chemicals are stored properly (ex. according to compatibility, not stored in fume hood). *In general, flammable chemicals should be stored away from oxidizing chemicals. Acids must be separated from caustic chemicals. Either distance or a barrier can be used for separation. Poisonous materials usually must be kept separate from acids. All chemicals must be stored and used away from any area used for eating or drinking. Chemicals with unusual properties should be stored separately from other chemicals. Storage areas should be labeled with DOT and NFPA labels.* | | | |
| 19 | Flammable liquids are stored away from ignition sources (burners, hotplates, electrical units, etc.). *If a container of flammable liquid failed, would the leaking liquid or vapor contact any item that could cause ignition?* | | | |
| 20 | All electrical wiring is free of fraying and cuts. *Electrical cords should not show signs of wear or breakage.* | | | |
| 21 | All electrical devices are grounded. *Three prong plugs should be used for all electrical items, except double insulated tools.* | | | |
| 22 | Extension cords are not used for permanent wiring. *Any fixed or permanent equipment should be hard wired into the power system. If the unit must be unpluggable, the outlet should be within reach without an extension cord. Computer systems may use a surge suppressing power strip to provide surge protection.* | | | |
| 23 | Controls that turn equipment on and off are labeled. *Both On and Off positions are identified. The equipment that is controlled by the switch is obvious, or the label includes the identification of the controlled equipment.* | | | |
| 24 | Electrical receptacles, switches, and controls are located so as not to be subject to liquid spills. *A Ground Fault Circuit Interrupter (GFCI) should be used on outlets within 6 feet of a water source.* | | | |
| 25 | Circuit breaker panels and electrical transformers are free of storage within 30 inches of the panel in laboratories and mechanical spaces. *Circuit breakers and other electrical disconnecting devices must have at least 30 inches of clearance to ensure immediate access if needed and to ensure electrician safety during maintenance.* | | | |

With EH&S = Laboratory environmental health and safety.

# REFERENCES

[1] Zakharyuta, Anastasia and Şen, Canhan and Avaz, Merve Senem and Akkaş, Tuğçe and Pürçüklü, Sibel and Baytekin Birkan, Tuğba and Gönül, Turgay and Yerdelen, Bilge and Cebeci, Fevzi Çakmak and İnce, Adnan) Sabanci University, Istanbul. (2016) *Laboratory Safety Handbook*. ISBN 978-605-9178-58-7.

[2] The University of Chicago *Environmental Health and Safety, Emergencies and Quick Response Guide* (2015, March). Retrieved from http://safety.uchicago.edu/guides/index.shtml

[3] University of Wisconsin-Madison Division of Facilities Planning and Management, Environment, *Lab Safety Guide, Chapter 5, Emergency Procedures* (2015, March). Retrieved from http://www.ehs.wisc.edu/chem/LabSafetyGuide-Chapter05.pdf

*Chapter 9*

# FIRST AID

Laboratories, even those well-equipped and controlled, are not immune to accidents and emergencies. When an occasional accident occurs, it is essential to know what to do and how to intervene, but first of all you need to know what you should not do. The behavior of the rescuer must be based on the speed of the decision and the rules of common sense.

General issues to be considered by the rescuer in the event of an accident are described below [1]:

P: PROTECT move the victim to a safe place.
I: INFORM call Emergency Services Number.
A: ASSIST
- Keep calm.
- Keep the victim warm.
- Do not move the victim if unaware of the seriousness of the situation.

## 9.1. INTOXICATIONS

Toxic elements may access the human body by:

- Ingestion;
- The respiratory tract;
- Skin contact.

The severity of the intoxication depends on the dose and toxicity of the product. For that reason, it is essential to know the product nature and what procedure is needed to remove it from the body [2].

A. *Intoxication by ingestion:*
- If the victim is unconscious:
1. Ask for urgent medical assistance.
2. Lean the victim forward and turn the victim's head to a side (recovery position).
3. Loosen the clothing and wrap the victim warm in a blanket.
4. DO NOT PROVOKE VOMIT. If the victim vomits, clear up the respiratory tract (cover your fingers with fabric to clean the victim's mouth).
5. Do not administer anything orally to an unconscious victim.
6. Do not attempt to neutralize the toxic product unless the Poison Control Centre indicates to do so.
- If the victim is conscious:
  If the toxic product is corrosive:
  1. Make the victim drink plenty of water.
  2. Take the victim urgently to a health center.
  3. DO NOT PROVOKE VOMIT.
  If the toxic product is not corrosive, take the victim to a health center.

B. *Intoxication through inhalation:*
This occurs when air containing toxic gas is breathed. In these cases somnolence and apathy arise, leading the victim to unconsciousness. Proceed as follows:
1. Immediately remove the injured person from the accident site and let the victim breath in non-stale air.

*First Aid* 155

2. Never attempt to save someone before verifying there is no risk to you and before asking for help previously.
3. If the injured person is conscious, apply artificial respiration at the first symptom of respiratory difficulty and take the victim to an emergency health centre.

C. *Intoxication through skin contact:*
   There are toxic substances that penetrate the body through skin or mucosa. Their toxicity depends on the type of product and the dose.
1. Remove contaminated clothing.
2. Wash the skin abundantly with water. Under no circumstance should the skin be rubbed.
   If the eyes are affected:
1. Rinse with water for 10-15 minutes to remove the product.
2. Take the patient to a medical center with the LABEL or MSDS of the product.

# 9.2. BURNS

A burn is tissue damage that results from chemical or electrical agents, or by radiation. Burns are classified into three different categories depending on the degree of the burn and the damage caused.

A. *Thermal burns:*
1. Run the burned area under cool tap water with low pressure for 10-15 minutes.
2. Size, depth and location of burned areas should be evaluated. If necessary, head to a health centre.
3. Burns should be covered with a wet gauze dressing or a clean piece of cloth.
4. Do not puncture blisters.

B. *Chemical burns: these are usually third degree burns and are caused by:*
Contact with chemical substances:
1. Rinse the skin with plenty of water for 10-15 minutes.
2. Wrap the burned area with a gauze dressing or a clean piece of cloth.
3. Remove clothing contaminated by the chemical, as long as it is not stuck to the skin.
4. Do not apply lotions or any other remedy to the burned area. Do not try to neutralise any chemical without consulting the Poison Control Centre.
5. Prevent any infection by covering the burned area with sterile gauze or clean cloth. The victim should be taken to a health centre.

Inhalation of chemicals (through the airways):
1. Remove the injured person from the accident site.
2. Wrap the burned area with sterile gauze dressing or cloth.
3. If not breathing, start artificial respiration.
4. If there is no pulse, start cardiopulmonary resuscitation (CPR).

C. *Electrical burns: these are usually third degree burns. There is generally no bleeding and they are painless.*

**Table 29. Types of burns**

| First degree | Only the outer layer of the skin is injured. | Skin is usually red and slightly painful. |
|---|---|---|
| Second degree | Epidermis and dermis layers are injured. | These can present blisters, pain and swelling. |
| Third degree | All layers of the skin are injured. | Dry skin. These present no pain due to nerve endings damage. |

Before starting first aid, flow of electricity should be interrupted by cutting the current. If this is not possible, proceed as follows:

1. The rescuer must be on a rubber or wooden surface.
2. The injured person should be removed from the source of electricity using a plastic or wooden object, as these do not conduct electricity.
3. Evaluate breathing and pulse. If there is no pulse, cardiopulmonary resuscitation should be started.

## 9.3. WOUNDS

Most wounds that occur in the chemistry laboratory are minor. For minor cuts, apply pressure to the wound with sterile gauze, wash with soap and water, and apply a sterile bandage. If the victim is bleeding badly, raise the bleeding part, if possible, and apply pressure to the wound with a piece of sterile gauze. While first aid is being given, someone else should notify the office clinic.

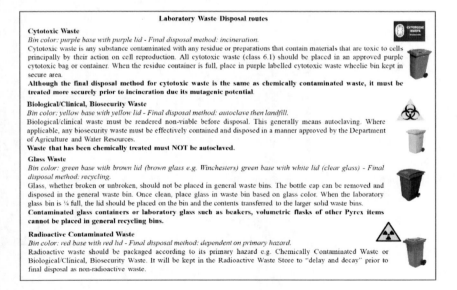

Figure 37. Required actions in the event of minor injuries.

Types of wounds:

A. *Abrasion:* it is produced when the skin is rubbed off against a rough surface. Generally, it doesn't bleed too much.
B. *Incised wound:* it is a cut in the skin caused by a sharp object. It is lineal and it bleeds from the divided vessels.
C. *Puncture wound:* it is caused by a sharp, pointed object that penetrates the skin, and that is why it is usually narrower but deeper. Cleaning and disinfecting wounds:
    1. Wash the wound using soap and running water.
    2. Disinfect the wound with hydrogen peroxide and sterile gauze.
    3. Apply an antiseptic solution to the wound (iodine) and place a dressing and a bandage to hold it in place.
    4. Cover incised wounds with gauze and head to a health centre in case it needs stitches.
    5. Help the bleeding in order to wash away any remaining bacteria.
    6. There is always a risk of tetanus infection, so everyone should be vaccinated against tetanus.

## 9.4. HEMORRHAGES

Blood circulates through the blood vessels throughout the whole body. When a blood vessel bursts, it bleeds and causes hemorrhage.

Hemorrhages or bleedings may be:

A. *External:* These occur when not only a blood vessel is broken, but also the skin, and so blood exits the body. They may be arterial, venous or capillary.
    1. Apply direct pressure at the bleeding point using a gauze or a freshly washed piece of cloth. If not available, do it with your hands using appropriate gloves.
    2. Elevate the injured area so as to reduce the blood pressure in it.

## First Aid

3. Apply direct compression over an artery against an underlying bone by using your fingers. This technique is used when direct pressure and elevation of the injured area fail to control the bleeding. In the event of nosebleed, the head should be tilted forward and the nose pinched together between the thumb and index fingers. Hold the pressure for a few minutes, and if the bleeding doesn't stop, place a dressing wetted in hydrogen peroxide into the nostril and go to the closest health centre.

B. *Internal:* These occur when blood doesn't exit the body, but it is collected under the skin (capillary bleeding), or in a cavity (venous or arterial bleeding).

Types of internal bleeding:

1. Capillary bleeding: trauma causes haematomas or tear of the tiny blood vessels under the skin. In the event of haematoma, apply ice in a bag against the skin, as cold temperatures cause blood vessels to tighten and thus the size of the haematoma is reduced.

2. Venous or arterial bleeding: it may be caused by abdominal trauma, such as falls from a height or car accidents. Symptoms like vomiting blood may be present.

## 9.5. BASIC CARDIOPULMONARY RESUSCITATION (CPR)

A. In the event of respiratory arrest:

1. Verify first the existence of thoracic movements. In the absence of these, proceed as follows:

2. Pinch the victim's nose closed with one hand, and keep the neck extended and the head tilted back with the other.

3. Inhale deeply, bring your mouth closer to the victim's half-open mouth and take a full breath that should bring the victim's chest up.

4. Move off and free the nose to allow air out.

## 160 — Maria Pia Gatto

B.  In the event of cardiac arrest (CPR):
1.  Locate the lower third of the victim's sternum.
2.  Place the heel of your hand on the sternum.
3.  Keep your fingers up off the chest wall. Place your other hand on top of the one that is in position and press down hard.
4.  Each compression should make the chest wall move inward about 5 cm.
5.  Use at least 100 *Chest Compressions per Minute* or 30 *Compressions to Every Two Breaths* (so as to pump blood in order to deliver oxygen to the brain).

## 9.6. ELECTRIC SHOCK SYMPTOMS

Injuries can be caused when electric current passes through body. The source may be natural or man-made. The danger depends on how high the voltage is, how the current traveled through the body, the person's overall health and how quickly the person is treated.

Call for emergency medical assistance if any of these symptoms occur:

- Unconsciousness.
- Cardiac arrest.
- Heart rhythm problems.
- Seizures.
- Respiratory failure.
- Muscle pain and contractions.
- Numbness and tingling.

What to do while waiting for medical help?

- Do not attempt to move the victim from source of current as touching the person may cause you to get shocked by the current as well.

*First Aid* 161

- Switch off the current source if possible otherwise, move the source using a nonconducting object like wooden stick.
- Prevent shock by laying the person down and, if possible, position the head slightly lower than the trunk, elevating the legs.
- Check for breathing.
- If there is no breathing, begin CPR.
- If the person is breathing, perform a physical examination.
- Treat for minor burns.

## 9.7. CHEMICAL SPILL

General procedure once the risk of injuries has been mitigated, the spill may be cleaned up and the area decontaminated using the following general procedures [3]:

1. Notify all personnel and supervisor in the vicinity of the spill, of any flammable, highly toxic or volatile material is spilled. Evacuate and post warnings in the area, if necessary, to isolate the area and prevent harmful exposure.
2. Provided the chemical spilled is not water reactive.
3. If clothing has become contaminated, remove and enter emergency shower, if eyes have been affected, flush eyes for 15 minutes.
4. Before responding to any spill the following information must be verified:
   a. Name of the chemical(s) involved.
   b. Approximate quantity.
   c. Hazards of the chemical (review SDS if available): Flammability: flash point; vapor pressure oToxicity – TLV Corrosiveness – pH
5. Perform clean-up procedures only if:
   a. The appropriate spill control material, equipment and protective clothing are available.
   b. Personnel are familiar with equipment and clean-up procedures.

c. More than one person is in the lab and available to participate. Work in teams. One person cleans the spill; the other should remain outside of the contaminated area and hand supplies to person cleaning.

d. There are no ignition sources present.

6. After reviewing the SDS and assessing the hazards posed by the spill, establish the appropriate clean-up procedure and supplies are on hand.

7. Determine the extent of evacuation required.

8. Gather the required equipment and materials.

9. Put on appropriate protective clothing. Minimum PPE includes lab coat, long loose fitting pants, and fully covering liquid resistant shoes. In addition, for performing a spill clean-up, medium or heavy duty rubber or nitrile gloves and safety goggles must be worn. Toxic, corrosive or irritating volatile materials will require the use of a respirator. Respirators must not be used without a model-specific fit test, and spill appropriate cartridges. A full-face respirator is the minimum requirement for volatile irritating, toxic or corrosive materials.

10. Use a spill control material (unreactive, neutral, compatible material) to make a 360 degree barrier around the spill and prevent it from seeping into a drain or under furniture or equipment.

11. Wait for any neutralizing/absorbent reactions to be complete, mix the spill control compound with the spill, and scoop the material into an impervious container.

12. Wash the affected area and PPE with an appropriate cleaning solution (soap and water).

13. Arrange for pick-up of the waste material.

# First Aid

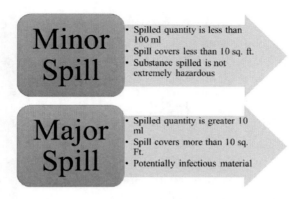

Figure 38. Types of spills.

A. *Cleaning a flammable solvent spill*

   *Note:* Never attempt to clean up a solvent spill if an ignition source is present

   1. Contain the spill to prevent it from spreading by using unreactive, neutral, compatible material (e.g., kitty litter, absorbent socks, or pads etc.) to create a barrier around the spill.
   2. Apply solvent absorbent (Spill X-S, Solusorb or equivalent product) from the perimeter inward, covering the total spill area.
   3. Mix thoroughly with plastic spatulas until material is dry and free flowing, and no evidence of free liquid remains.
   4. Transfer the absorbed solvent to an appropriate disposal container that is not soluble in the solvent, and seal the container.
   5. Contact the ESF for directions concerning disposal of the container and its contents. Procedure is complete, method of disposal and expectation of general housekeeping is detailed.

B. *Cleaning an acid spill*

   Except hydrofluoric acid and perchloric acid – these acids require specialized handling practices

1. Contain the spill to prevent it from spreading by using an unreactive, neutral, compatible material (e.g., kitty litter, absorbent socks, or pads etc.) to create a barrier around the spill.
2. Apply acid neutralizer (Spill X-A, Neutrasorb or equivalent product) gently to the spill.
3. Carefully mix with a plastic spatula or other tool, working towards the spill center to minimize spread.
4. When foaming subsides, check pH with pH paper.
5. If pH is less than pH 6, add more neutralizer to any free acid and repeat step 3 and 4; if spill pH is greater than pH 8 add a caustic neutralizer (Spill X-C or equivalent). The target range is pH 6 to 8.
6. When the spill has been sufficiently neutralized, pick up treated material with scoops, dust pan, broom, and transfer to a disposal container.
7. Seal and label container.
8. Decontaminate and wash spill site surfaces with soapy water and wet sponge.
9. Contact the ESF for directions concerning disposal of the bag and its contents.

C. *Caustic*
1. Contain the spill to prevent it from spreading by using unreactive, neutral, compatible material (eg. kitty litter, absorbent socks, or pads etc.) to create a barrier around the spill.
2. Gently apply neutralizer for caustics (Spill X-C, Neutracit-2 or equivalent product) to the spill, working inwards.
3. Carefully mix with a plastic spatula or other tool;
4. When foaming subsides, check pH with pH paper.
5. If pH is greater than 8, add more neutralizer to any free base and repeat step 3 and 4; if pH is less than 6, add acid neutralizer and repeat step 3 and 4. The target range is pH 6 to 8.

## First Aid

6. When the spill has been sufficiently neutralized, pick up treated material with scoops, dust pan, broom and transfer to a disposal container.
7. Seal and label container.
8. Decontaminate and wash spill area surfaces with water and wet sponge.
9. Check with the ESF for directions concerning disposal of the bag and contents.

# 9.8. CHECKLIST FOR THE PREVENTION OF ACCIDENTS IN LABORATORIES

The purpose of this checklist is to promote safety awareness and to encourage safe work practices in the laboratory. This checklist should serve as a reminder of things you can do to work more safely. The checklist comprises of a series of questions to determine whether or not the laboratory conforms with the regulation and methods of best practice [4]. All questions must be answered with Yes or No. In the latter case a corrective action is needed.

### Table 30. General safety and lab students/workers risk preception

|  | Yes | No |
|---|---|---|
| **General Laboratory Safety** | | |
| Are staff and/or students adequately trained and supervised to ensure safe work procedures? | | |
| Are new and young workers appropriately supervised and informed? | | |
| Are hazard warning signs posted on the door indicating any hazards that may be present (e.g., biological, radioactive materials, or high noise emitting equipment)? | | |
| Is there a door that can be closed to keep visitors out of the lab while work with the dangerous substances or biological agents is in progress? | | |
| Where required, is access to the lab restricted to authorised persons only? | | |
| Are storage areas (rooms, refrigerators, freezers, cupboards) where infectious and/or toxic materials are kept labelled accordingly? | | |
| Are all chemicals clearly labelled, including hazard symbols? | | |

# Maria Pia Gatto

## Table 30. (Continued)

| | Yes | No |
|---|---|---|
| Are all biomaterials clearly labelled, including the hazard symbol where appropriate? | | |
| Are glass bottles stored where they cannot be knocked or kicked over? | | |
| Are all pressure vessels (including pressurised liquid nitrogen dewars) periodically inspected and certified? | | |
| Is the safe working pressure clearly marked on all pressure vessels? | | |
| Are all pressurised gas cylinders properly secured by restraining chains, bench clamps or similar? | | |
| Are gas cylinders sited away from doors or escape routes? | | |
| Is all portable electrical equipment periodically tested and labelled with the date of test? | | |
| Is the use of laboratory equipment (such as electrophoresis for instance) covered by safety instructions? | | |
| Are the power supply leads to electrophoresis equipment covered? | | |
| Do all centrifuges have interlocked lids? | | |
| If no, is a suitable warning sign affixed to the centrifuge lid? | | |
| Are floors clean? | | |
| Are floor coverings intact and non-slip? | | |
| Are passageways clear of tripping hazards (cables, stock, waste, etc.)? | | |
| Are work surfaces easily cleaned and decontaminated after use? | | |
| Is lighting adequate and in working order? | | |
| Are noise levels acceptable? (you do not have to raise your voice to talk to people at your workplace) | | |
| Is all portable electrical equipment periodically tested and labelled with the date of test? | | |
| Is the use of laboratory equipment (such as electrophoresis for instance) covered by safety instructions? | | |
| Are the power supply leads to electrophoresis equipment covered? | | |
| Do all centrifuges have interlocked lids? | | |
| If no, is a suitable warning sign affixed to the centrifuge lid? | | |
| Are floors clean? | | |
| Are floor coverings intact and non-slip? | | |
| Are passageways clear of tripping hazards (cables, stock, waste, etc.)? | | |
| Are work surfaces easily cleaned and decontaminated after use? | | |
| Is lighting adequate and in working order? | | |
| Are noise levels acceptable? (you do not have to raise your voice to talk to people at your workplace) | | |
| **Information for Workers** | | |
| Are staff and/or students adequately trained and supervised to ensure safe work procedures? | | |
| Are new and young workers appropriately supervised and informed? | | |
| Are hazard warning signs posted on the door indicating any hazards that may be present (e.g., biological, radioactive materials, or high noise emitting equipment)? | | |

# First Aid

|  | Yes | No |
|---|---|---|
| Is there a door that can be closed to keep visitors out of the lab while work with the dangerous substances or biological agents is in progress? | | |
| Where required, is access to the lab restricted to authorised persons only? | | |
| Are storage areas (rooms, refrigerators, freezers, cupboards) where infectious and/or toxic materials are kept labelled accordingly? | | |
| Are all chemicals clearly labelled, including hazard symbols? | | |
| Are all biomaterials clearly labelled, including the hazard symbol where appropriate? | | |
| Are glass bottles stored where they cannot be knocked or kicked over? | | |
| Are all pressure vessels (including pressurised liquid nitrogen dewars) periodically inspected and certified? | | |
| Is the safe working pressure clearly marked on all pressure vessels? | | |
| Are all pressurised gas cylinders properly secured by restraining chains, bench clamps or similar? | | |
| Are gas cylinders sited away from doors or escape routes? | | |
| Is all portable electrical equipment periodically tested and labelled with the date of test? | | |
| Is the use of laboratory equipment (such as electrophoresis for instance) covered by safety instructions? | | |
| Are the power supply leads to electrophoresis equipment covered? | | |
| Do all centrifuges have interlocked lids? | | |
| If no, is a suitable warning sign affixed to the centrifuge lid? | | |
| Are floors clean? | | |
| Are floor coverings intact and non-slip? | | |
| Are passageways clear of tripping hazards (cables, stock, waste, etc.)? | | |
| Are work surfaces easily cleaned and decontaminated after use? | | |
| Is lighting adequate and in working order? | | |
| Are noise levels acceptable? (you do not have to raise your voice to talk to people at your workplace) | | |
| Is all portable electrical equipment periodically tested and labelled with the date of test? | | |
| Is the use of laboratory equipment (such as electrophoresis for instance) covered by safety instructions? | | |
| Are the power supply leads to electrophoresis equipment covered? | | |
| Do all centrifuges have interlocked lids? | | |
| If no, is a suitable warning sign affixed to the centrifuge lid? | | |
| Are floors clean? | | |
| Are floor coverings intact and non-slip? | | |
| Are passageways clear of tripping hazards (cables, stock, waste, etc.)? | | |
| Are work surfaces easily cleaned and decontaminated after use? | | |
| Is lighting adequate and in working order? | | |
| Are noise levels acceptable? (you do not have to raise your voice to talk to people at your workplace) | | |

## REFERENCES

[1] University of Wisconsin-Madison Division of Facilities Planning and Management, Environment, *Lab Safety Guide, Chapter 5, Emergency Procedures* (2015, March). Retrieved from http://www.ehs.wisc.edu/chem/LabSafetyGuide-Chapter05.pdf

[2] Furr, A. K. (2000). *CRC Handbook of Laboratory Safety*. Care and Use of Electrical Systems (5th ed.) (pp. 328-336). United States of America: CRC Press LLC.

[3] The University of Texas at Austin Environmental Health and Safety, *Emergency Instructions for Labs* (2015, March). Retrieved from https://www.utexas.edu/safety/ehs/lab/manual/2_emergency.html

[4] *Checklist for the Prevention of Accidents in Laboratories* European. OSHA e-Facts 20 Agency for Safety and Health at Work - http://osha.europa.eu

*Chapter 10*

# LABORATORY SAFETY STANDARDS

Laboratory Safety Standards are used in clinical and chemical laboratories, testing laboratories, as well as research and development laboratories in various areas and educational facilities. The laboratory's safety standards relate to clothing and equipment, as well as laboratory procedures and designs used. Find the standards from International Organization of Standardization (ISO); International Safety Equipment Association (ISEA); ASTM International (ASTM); International Electrotechnical Commission (IEC); British Standards Institution (BSI); Laser Institute of America (LIA); Clinical and Laboratory Standards Institute (CLSI); Standards Australia (SAI); and American Industrial Hygiene Association (AIHA).

There are many new laboratory safety standards introduced every year. Many important laboratory safety standards have been categorized below.

## 10.1. PROTECTIVE CLOTHING

*ISO 6529:2013*

Protective clothing - Protection against chemicals - Determination of resistance of protective clothing materials to permeation by liquids and gases

## ISO 16603:2004

Clothing for protection against contact with blood and body fluids - Determination of the resistance of protective clothing materials to penetration by blood and body fluids - Test method using synthetic blood

## ISO 16604:2004

Clothing for protection against contact with blood and body fluids - Determination of resistance of protective clothing materials to penetration by blood-borne pathogens - Test method using Phi-X 174 bacteriophage

## ISO 13994:2005

Clothing for protection against liquid chemicals - Determination of the resistance of protective clothing materials to penetration by liquids under pressure

## ANSI/ISEA 101-2014

American National Standard for Limited-Use and Disposable Coveralls - Size and Labeling Requirements

# 10.2. HAND/EYE PROTECTION

## ANSI/ISEA 105-2016

American National Standard for Hand Protection Classification

## ANSI/ISEA Z358.1-2014

American National Standard for Emergency Eyewash and Shower Equipment

## ANSI/ISEA Z87.1-2015

American National Standard for Occupational and Educational Personal Eye and Face Protection Devices

*Laboratory Safety Standards* 171

*ASTM E2011-13*
Standard Test Method for Evaluation of Hygienic Handwash and Handrub
  Formulations for Virus-Eliminating Activity Using the Entire Hand


## 10.3. EQUIPMENT SAFETY

*IEC 61010-031 Ed. 2.0 b:2015*
Safety requirements for electrical equipment for measurement, control and
  laboratory use - Part 031: Safety requirements for hand-held probe
  assemblies for electrical measurement and test

*S+ IEC 61010-031 Ed. 2.0 en:2015 (Redline version)*
Safety requirements for electrical equipment for measurement, control and
  laboratory use - Part 031: Safety requirements for hand-held probe
  assemblies for electrical measurement and test

*IEC 61010-1 Ed. 3.0 b:2010*
"Safety requirements for electrical equipment for measurement, control, and
  laboratory use - Part 1: General requirements"

*IEC 61010-2-010 Ed. 3.0 b:2014*
Safety requirements for electrical equipment for measurement, control and
  laboratory use - Part 2-010: Particular requirements for laboratory
  equipment for the heating of materials

*IEC 61010-2-020 Ed. 3.0 b:2016*
Safety requirements for electrical equipment for measurement, control, and
  laboratory use - Part 2-020: Particular requirements for laboratory
  centrifuges

*IEC 61010-2-030 Ed. 2.0 b:2017*

Safety requirements for electrical equipment for measurement, control, and laboratory use - Part 2-030: Particular requirements for equipment having testing or measuring circuits

*IEC 61010-2-032 Ed. 3.0 b:2012*

Safety requirements for electrical equipment for measurement, control and laboratory use - Part 2-032: Particular requirements for hand-held and hand-manipulated current sensors for electrical test and measurement

*IEC 61010-2-033 Ed. 1.0 b:2012*

"Safety requirements for electrical equipment for measurement, control, and laboratory use - Part 2-033: Particular requirements for hand-held multimeters and other meters, for domestic and professional use, capable of measuring mains voltage"

*IEC 61010-2-040 Ed. 2.0 b:2015*

Safety requirements for electrical equipment for measurement, control, and laboratory use - Part 2-040: Particular requirements for sterilizers and washer-disinfectors used to treat medical materials

*IEC 61010-2-081 Ed. 2.0 b:2015*

Safety requirements for electrical equipment for measurement, control and laboratory use - Part 2-081: Particular requirements for automatic and semi-automatic laboratory equipment for analysis and other purposes

*IEC 61010-2-091 Ed. 1.0 b:2012*

Safety requirements for electrical equipment for measurement, control and laboratory use - Part 2-091: Particular requirements for cabinet X-ray systems

*IEC 61010-2-201 Ed. 2.0 b:2017*

Safety requirements for electrical equipment for measurement, control, and laboratory use - Part 2-201: Particular requirements for control equipment

*IEC 61326-3-1 Ed. 2.0 b:2017*

Electrical equipment for measurement, control and laboratory use - EMC requirements – Part 3-1: Immunity requirements for safety-related systems and for equipment intended to perform safety-related functions (functional safety) – General industrial applications

*IEC 61326-3-2 Ed. 2.0 b:2017*

Electrical equipment for measurement, control and laboratory use - EMC requirements - Part 3-2: Immunity requirements for safety-related systems and for equipment intended to perform safety-related functions (functional safety) - Industrial applications with specified electromagnetic environment

*BS EN 13150:2001*

Workbenches for laboratories. Dimensions, safety requirements and test methods (British Standard)

## 10.4. LASER SAFETY

*IEC 60825* - Safety of Laser Products Package

*IEC 60825* - Safety of Laser Products Package

*ANSI Z136.1-2014*

American National Standard for Safe Use of Lasers

*ANSI Z136.2-2012*
American National Standard for Safe Use of Optical Fiber Communication Systems Utilizing Laser Diode and LED Sources

*ANSI Z136.4-2010*
American National Standard Recommended Practice for Laser Safety Measurements for Hazard Evaluation

*ANSI Z136.5- 2009*
American National Standard for Safe Use of Lasers in Educational Institutions

*ANSI Z136.6-2015*
Safe Use of Lasers Outdoors

*ANSI Z136.8-2012*
American National Standard for Safe Use of Lasers in Research, Development, or Testing

*ANSI Z136.9-2013*
American National Standard for Safe Use of Lasers in Manufacturing Environments

## 10.5. LABORATORY TESTING AND MANAGEMENT

*ISO/IEC 17025:2017*
General requirements for the competence of testing and calibration laboratories

*ISO 10012 and ISO/IEC 17025*
Laboratories Measurement Management System Requirements Package

*ISO 10012 and ISO/IEC 17025*
Laboratories Measurement Management System Requirements Package

*ISO/IEC 17025 and ISO 15189*
Competence Testing and Calibration of Medical Laboratories Package

*ISO/IEC 17025 and ISO 15189*
Competence Testing and Calibration of Medical Laboratories Package

*ISO/IEC 17043 / ISO/IEC 17025 / ISO/IEC 17000*
Competence and Proficiency Testing Package

*ISO/IEC 17043 / ISO/IEC 17025 / ISO/IEC 17000*
Competence and Proficiency Testing Package

## 10.6. LABORATORY DESIGN AND PROCEDURES

*ANSI/ASSE Z9.1-2016*
Ventilation and Control of Airborne Contaminants During Open-Surface
Tank Operations

*ANSI/ASSE Z9.3-2017*
Spray Finishing Operations: Safety Code for Design, Construction and
Ventilation

*ANSI/AIHA/ASSE Z9.4-2011*
Abrasive-Blasting Operations - Ventilation and Safe Practices for Fixed
Location Enclosures

*ANSI/AIHA/ASSE Z9.5-2012*
Laboratory Ventilation

*ANSI/AIHA/ASSE Z9.6-2008*
Exhaust Systems for Grinding, Polishing, and Buffing

*ANSI/AIHA/ASSE Z9.7-2007*
Recirculation of Air from Industrial Process Exhaust Systems

*ANSI/AIHA/ASSE Z9.9-2010*
Portable Ventilation Systems

*ANSI/ASSE Z9.10-2017*
Fundamentals Governing the Design and Operation of Dilution Ventilation Systems in Industrial Occupancies

*ANSI/ASSE Z9.11-2016*
Laboratory Decommissioning

*ISO 8573* - Compressed Air Package
*ISO 8573-1, ISO 8573-2, ISO 8573-3, ISO 8573-4, ISO 8573-5, ISO 8573-6, ISO 8573-7, ISO 8573-8 and ISO 8573-9*

*ASTM E2093-12(2016)*
Standard Guide for Optimizing, Controlling and Assessing Test Method Uncertainties from Multiple Workstations in the Same Laboratory Organization

## 10.7. OTHER

*CGA P-1_OSHA*
Cited by OSHA, 5th edition "Safe Handling of Compressed Gases"

*ISO 11625:2007*
Gas cylinders - Safe handling

*NFPA 45-2011*

*NFPA 45*
Standard on Fire Protection for Laboratories Using Chemicals, 2011 Edition

*Chapter 11*

# GLOSSARY

**A**

*absorbed dose (in toxicology):* The amount of a chemical absorbed into the body or into organs and tissues of interest (WHO, 1978a).

*absorbed dose (in radiation):* The energy imparted to matter in a suitably small element of volume by ionizing radiation divided by the mass of that element of volume (ISO, 1972). The SI unit for absorbed dose is joule per kilogram (J kgW-1) and its special name is gray (Gy) (ISO, 1972).

*authorized person:* A person approved or assigned by the employer to perform a specific type of duty or duties or to be at a specific location or locations at the jobsite. See *designated person.*

*acceptable daily intake:* This is an estimate of the amount of substance in the food that can be ingested daily over a lifetime by humans without appreciable health risk. The concept of the ADI has been developed principally by WHO and FAO and is relevant to chemicals such as additives to foodstuffs, residues of pesticides and veterinary drugs in foods. ADIs are derived from laboratory toxicity data, and from human experiences of such chemicals when this is available, and incorporate the safety factor.

180 *Glossary*

*acceptable daily intake (pesticide residues):* The acceptable daily intake of a chemical is the daily intake which, during an entire life time, appears to be without appreciable risk to the health of the consumer on the basis of all the known facts at the time when a toxicological assessment is carried out. It is expressed in milligrams of the chemical per kilogram of body weight (Vettorazzi, 1980).

*acceptable daily intake not specified:* An ADI without an explicit indication of the upper limit of intake may be assigned to substances of very low toxicity, especially those that are food constituents or that may be considered as foods or normal metabolites in man.

*acceptable risk:* This concept relates to the probability of suffering disease or injury that will be tolerated by an individual, group or society. Acceptability of risk depends on the scientific data, social, economic and political factors, and on the perceived benefits arising from the a chemical or process.

*accident prevention:* A set of precautionary, measures taken to avoid possible bodily harm.

*accumulation:* Successive additions of a substance to a target organism, or organ, or to part of the environment, resulting in an increasing quantity or concentration of the substance in the organism, organ, or environment.

*action level:* (i) The level of a pollutant at which specified emergency countermeasures, such as the seizure and destruction of contaminated materials, evacuation of the local population or closing down the sources of pollution, are to be taken (UN, 1972); (ii) the level at which some kind of preventive action (not necessarily of an emergency nature) is to be taken; (iii) a level of exposure of workers to airborne harmful substances in workrooms to be determined by the competent authority; it is distinctly below the exposure limit and consequently such exposures below the action level do not usually necessitate application of all the preventive measures, especially of a medical nature, foreseen for exposures exceeding the action level. This level may lie between a third and a half of the exposure limit (ILO, 1977).

# Glossary 181

*acute effects:* Effects that occur rapidly following exposure and are of short duration (WHO, 1979).

*acute toxicity:* The adverse effects occurring within a short time of administration of a single dose or multiple doses given within 24 hours (Hagan, 1959).

*acute toxicity test:* An experimental animal study in which the adverse effects occur in a short time (from 1-7 days) following the administration of a single or multiple doses of a chemical. The most frequently used acute toxicity test involves determination of the median lethal dose (LD50) of the compound. The LD50 has been defined as ``a statistically derived expression of a single administered dose of a material that can be expected to kill 50% of the animals" (WHO, 1978a).

*additive effect:* An additive effect is the overall consequence which is the result of two chemicals acting together and which is the simple sum of the effects of the chemicals acting independently. See also antagonistic effect, synergistic effect.

*adsorption:* A process whereby one or more components of an interfacial layer between two bulk phases are either enriched or depleted (IUPAC, 1972).

*adverse effect:* This is abnormal, undesirable or harmful effect to an organism, indicated by some result such as mortality, altered food consumption, altered body and organ weights,altered enzyme levels or visible (pathological) change. An effect may be classed as adverse effect if it causes functional or anatomical damage, causes irreversible changes or increases the susceptibility of the organism to other chemical or biological stress. A non-adverse effect will usually be reversed when exposure to the chemical ceases.

*allergen:* This descriptor may be used to any substance which produces an allergic reaction.

*allergy:* A broad term applied to disease symptoms following exposure to a previously encountered substance (allergen), often one which would otherwise be classified as harmless. Essentially it is a malfunction of the immune system. See sensitization.

182                                   *Glossary*

*antagonistic effect*: This is the consequence of one chemical (or a group of chemicals) interacting: the situation in which the combined effect of two or more chemicals is less than the simple sum of their independent effects. In bioassay, the term may be used to refer to the situation when a specified response is produced by exposure to either of two factors but not by exposure to both together (Last, 1983).

## B

*bioaccumulation*: The process by which the amount of a substance in a living organism (or its parts) increases with time (WHO, 1979).

*biochemical mechanism*: This is the general term for any chemical reaction or series of reactions, usually enzyme catalysed, which produces a given physiological effect in a living organism.

*biological assessment of exposure*: Exposure to chemicals may be assessed by the analysis of specimens taken in the environment (air, water, food, etc.) or of specimens of biological material. Most often, urine and blood are analyzed, but other materials such as expired air, faeces, saliva, bile, hair, and biopsy or autopsy material are sometimes analyzed. In these samples, the content of the xenobiotic(s) or its metabolite(s) is determined and, on this basis, the exposure level (concentration in the air, absorbed amount of the substance) or the probability of health impairment due to exposure is derived. Biochemical changes in the components of an organism can also be used for this purpose (e.g., changes in enzyme activity or in the excretion of metabolic intermediates) if they show a relationship to the exposure (WHO, 1979).

*biological monitoring*: The periodic examination of biological specimens (in accordance with the definition of monitoring). It is usually applied to exposure monitoring but can also apply to effect monitoring (WHO, 1979).

# Glossary

## C

*cancer*: Cancer is a disease which results from the development of a malignant tumour and its spread into the surrounding tissues. See tumour.

*carcinogen*: An agent, chemical, physical or biological, that can act on living tissue in such a way as to cause a malignant neoplasm (WHO, 1980).

*carcinogenesis*: The induction by chemical, physical, or biological agents, of neoplasms that are usually not observed, an earlier induction of neoplasms that are usually observed, and/or the induction of more neoplasms than are usually found although fundamental differences in the mechanisms may be involved (IARC, 1977).

*chromosomal aberration*: Any abnormality of chromosome number or structure may be described as an aberration.

*chronic effects*: Effects that develop slowly and have a long duration. They are often, but not always, irreversible. Some irreversible effects may appear a long time after the chemical substance was present in the sensitive tissue. For such delayed or late effects, the latent period (or the ``time to occurrence" of an observable effect) may be very long, particularly if the level of exposure is low (WHO, 1979).

*chronic toxicity test*: A study in which animals are observed during the whole life span (or the major part of the life span) and in which exposure to the test material takes place over the whole observation time or a substantial part thereof. The term ``long-term toxicity study" is sometimes used as a synonym for ``chronic toxicity study" and sometimes to signify a study that falls in between subacute (short- term toxicity studies) and chronic toxicity studies (WHO, 1978a).

*ceiling value (CV)*: The maximum permissible airborne concentration of a potentially toxic substance and is a concentration that should never be exceeded in the breathing zone.

*C.I.H.*: Certified Industrial Hygienist

*concentration*: A general term referring to the quantity of a material or substance contained in unit quantity of a given medium. When the term

184                                    *Glossary*

concentration is used without further qualification, it now means amount of substance concentration (WHO, 1979).

*conditional acceptable daily intake*: A conditional acceptable daily intake is one that is established for a pesticide in order to limit its use to those situations where no satisfactory substitutes are avail- able. This definition will be the subject of further discussion. The allocation of conditional ADIs for intentional food additives has been superseded (Vettorazzi, 1980).

*contaminant*: In some contexts (e.g., in relation to gas cleaning equipment), used as a synonym for pollutant (ISO, 1979).

*control limit*: A regulatory value applied to the airborne concentration in the workplace of a potentially toxic substance which is judged to be "reasonably practicable" for the whole spectrum of work activities and which must not normally be exceeded.

*competent person*: One who is capable of identifying existing and predictable hazards in the surroundings, or working conditions which are unsanitary, hazardous, or dangerous to employees, and who has authorization to take prompt corrective measures to eliminate them.

*corrective actions*: A change implemented to address a weakness identified in a management system. Normally corrective actions are instigated in repose to a customer complaint.

*corrosive of tissue*: The descriptor applied to any substance which destroys tissues on direct contact.

*critical group*: That part of the target population most in need of protection (WHO, 1979).

*critical organ (critical tissue)(in toxicology)*: The particular organ that first attains the critical concentration (of metal) under specified circumstances of exposure and for a given population (Task Group on Metal Toxicology, 1976).

*critical organ(in radiation biology)*: The organ whose damage (by radiation) results in the greatest injury to the individual (or his descendants). The injury may result from inherent radiosensitivity or indispensability of the organ, or from high dose, or from a combination of all three (ICRP, 1965).

# Glossary

185

*crude death rate*: See death rate.

*cumulative effect*: Occurs when repeated doses of a toxic substance or harmful radiation summate to give an enhanced effect (WHO, 1979).

*cumulative incidence ratio*: The ratio of the cumulative incidence rate in the exposed to the cumulative incidence rate in the unexposed (Last, 1983).

*cyanosis*: The pathological condition where there is an excessive concentration of reduced haemoglobin in the blood. This results in blue appearance of the skin, especially on the face and extremities, indicating the lack of sufficient oxygen in arterial blood.

*cytotoxic*: The adjective applied to anything that is harmful to the cell structure and function and ultimately causing cell death.

## D

*death rate*: An estimate of the proportion of a population that dies during a specified period. The numerator is the number of persons dying during the period; the denominator is the size of the population, usually estimated as the mid-year population.

*dermatitis*: Inflammation of the skin.

*distribution*: This is a general term for the dispersal of a applied substance and its derivatives throughout an organism or environmental system.

*dose:* The amount of a chemical administered to an organism (WHO, 1978).

*dose exposure-response relationship*: The relationship between administered dose or exposure and the biological change in organisms. It may be expressed as the severity of an effect in one organism (or part of an organism) or as the proportion of a population exposed to a chemical that shows a specific reaction (WHO, 1979).

## E

*ecotoxicology:* The effects of chemical agents on the environment, including, in addition to effects on man, adverse events that take place in the general ecosystem. It is not necessarily related primarily to human health (WHO, 1979).

186 *Glossary*

*effect*: A biological change in an organism, organ, or tissue (WHO, 1979).

elimination (in metabolism): The expelling of a substance or other material from the body (or a defined part thereof), usually by a process of extrusion or exclusion but sometimes through metabolic transformation (WHO, 1979).

*embryotoxicity*: The potential of a substance to induce adverse effects in progeny in the first period of pregnancy between conception and the fetal stage (UNEP/IRPTC, 1982).

*emission*: The giving off of environmental pollutants from various sources (WHO, 1979).

*emission or exposure control*: The technical and administrative procedures applied for the reduction or elimination of emissions from the source or of exposure to the target (WHO, 1988).

*environmental quality standard (EQS)*: This regulatory value defines the maximum concentration of a potentially toxic substance which can be allowed in an environmental compartment, usually air or water, over a defined period. Synonym: ambient standard. See limit values.

*enzymes*: Proteins which act as highly selective catalyst. This permits reactions in living cells to take place rapidly under physiological conditions. Enzymes are also used in the industry, for example as additives in the detergents.

*excretion*: The discharge or elimination of an absorbed or endogenous substance, or of a waste product, via some tissue of the body and its appearance in urine, faeces, or other products normally leaving the body. Excretion of chemical compounds from the body occurs mainly through the kidney and the gut. For volatile compounds, however, elimination by exhalation may be important. Excretion via perspiration and through hair and nails may also be of importance under special circumstances. Excretion via the gastrointestinal tract may take place by various routes such as the bile, the shedding of intestinal cells, and transport through the intestinal mucosa (WHO, 1979).

*excretion rate*: The amount or proportion of a substance that is excreted per unit time. It should be noted that according to this definition excretion does not include the passing of a substance through the intestine without

# Glossary 187

absorption. When discussing the total amount of a substance in faeces (including the unabsorbed part), it is preferable to speak about faecal content of substance (Task Group on Metal Accumulation, 1973).

*exposed group (in epidemiology)*: A group whose members have been exposed to a supposed cause of a disease or health state of interest, or possess a characteristic that is a determinant of the health outcome of interest. The abbreviated term ``the exposed" is sometimes used.

*exposed or non-exposed*: Qualitative terms defining the existence of or lack of a hazard in the environment of individuals (WHO, 1988).

*exposure*: The amount of an environmental agent that has reached the individual (external dose) or has been absorbed into the individual (internal dose, absorbed dose) (WHO, 1979).

*exposure assessment*: The quantification of the amount of exposure to a hazard for an individual or group (WHO, 1979).

*exposure control*: see emission or exposure control.

*exposure limit*: A general term implying the level of exposure that should not be exceeded (WHO, 1979).

*extrapolation*: The calculation, based on quantitative observations in exposed test species, of predicted dose-effect and dose-response relationships for a chemical in humans and other environmental biota (WHO, 1979).

## F

*fetus (foetus)*: In medicine, this term is applied to the young of mammals when fully developed in the womb. In humans, this stage is reached after about 3 months of pregnancy. Prior to this, the developing mammal is in embryo stage.

*fire precautions*: The measures taken and the fire protection features provided in a building (e.g., design, systems, equipment and procedures) to minimise the risk to the occupants from the outbreak of fire.

*fire prevention*: The concept of preventing outbreaks of fire, of reducing the risk of fire spreading and of avoiding danger to persons and property from fire.

188 *Glossary*

*first aid*: The skilled application of accepted principles of treatment on the occurrence of an accident or in the case of sudden illness, using facilities or materials available at the time.

*follow-up study (synonym: cohort study)*: A study in which individuals or populations, selected on the basis of whether they have been exposed to risk, received a specified preventive or therapeutic procedure, or possess a certain characteristic, are followed to assess the outcome of exposure, the procedure, or effect of the characteristic, e.g., occurrence of disease (from Last, 1988).

*fungicide*: A chemical used to kill fungi. See pesticide.

## G

*graded effect*: An effect that can usually be measured on a graded scale of intensity or severity and its magnitude related directly to the dose (WHO, 1978a).

*guideline level*: A guideline level is the maximum concentration of a pesticide residue that might occur after the officially recommended or authorized use of a pesticide for which no acceptable daily intake or temporary acceptable daily intake is established and that need not to be exceeded if good practices are followed. It is expressed in milligrams for the residue per kilogram of food (WHO, 1976).

## H

*harm*: Injury to or death of persons, or damage.

*hazard*: A source of danger: a qualitative term expressing the potential that an environmental agent can harm health (WHO, 1988).

*hazard identification*: The identification of the substance of concern, its adverse effects, target populations, and conditions of exposure (WHO, 1988).

*health*: A state of complete physical, mental and social well-being, and not merely the absence of disease or infirmity (WHO, 1978b).

## Glossary 189

*healthy worker effect*: A phenomenon observed initially in studies of occupational diseases: workers usually exhibit lower overall death rates than the general population, due to the fact that the severely ill and disabled are ordinarily excluded from employment. Death rates in the general population may be inappropriate for comparison if this effect is not taken into account (Last, 1983).

*hepatotoxic*: The adjective applied to anything which is harmful to the liver.

*herbicide*: The descriptor applied to a chemical used to kill plants. See pesticide.

*HSE*: Health and Safety Executive. A statutory body, established under the Health and Safety at Work etc. Act 1974 (HSWA). It is an enforcing authority working in support of the HSC. Local authorities are also enforcing authorities under HSWA. See: http:// www.hse.gov.uk

*hypersensitivity*: See allergy.

## I

*immune response*: The immune response is a general reaction of the body to substances that are foreign or treated as foreign. It may take various forms: antibody production,cell-mediated immunity, immunological tolerance, or allergy.

*incidence*: The number of instances of illness commencing, or of persons falling ill, during a given period in a specific population. Incidence is usually expressed as a rate, the denominator being the average number of persons in the specified population during a defined period or the estimated number of persons at the mid-point of that period. The basic distinction between incidence and prevalenceis that whereas incidence refers only to new cases, prevalence refers to all cases, irrespective of whether they are new or old. When the terms incidence and prevalence are used, it should be stated clearly whether the data represent the numbers of instances of the disease recorded or the numbers of persons ill (WHO, 1966).

*incidence rate*: The rate at which new events occur in a population. The numerator is the number of new events that occur in a defined period;

190 *Glossary*

the denominator is the population at risk of experiencing the event during this period, sometimes expressed as person-time.

*insecticide*: A chemical used to kill insects. See pesticides.

*intake*: The amount of a substance or material that is taken into the body, regardless of whether or not it is absorbed. The daily intake may be expressed as the amount taken in by a particular exposure route, e.g., ingestion or inhalation. The daily intake from food is the total amount of a given substance taken in during one day through the consumption of food. The daily intake by inhalation is calculated by multiplying the concentration of the substance (or agent) in air by the total amount of air inhaled during one day (24 hours). The total daily intake is the sum of the daily intake by an individual from food, drinking-water, and inhaled air (WHO, 1979).

*in vitro*: A term applied to any study carried out in isolation from the living organism in an experimental system (`in a test tube').

*in vivo*: The term used in contrast with `in vitro' describing any study carried out within the living organism.

*irritant*: Applied to any substance causing inflammation following immediate, prolonged or repeated contact with skin or mucous membranes.

**J**

*JIGSR:* Joint Inter-Institutional Group on Safety and Risk.

*JSA:* Job Safety Analysis – is a common term used in construction or jobsite environment by Field Personal or Project Mangers. An excellent example of which General Contractors and Sub contractors must comply to for a NJSCC project.

**L**

*latent period (synonym: latency):* Delay between exposure to a disease-causing agent and the appearance of manifestations of the disease. After exposure to ionizing radiation, for instance, there is a latent period of

# Glossary 191

five years, on average, before development of leukemia, and more than 20 years before development of certain other malignant conditions. The term ``latent period'' is often used as synonym with ``induction period'', that is, the period between exposure to a disease-causing agent and the appearance of manifestations of the disease. It has also been defined as the period from disease initiation to disease detection (Last, 1988).

*LC50:* This abbreviation is used for the exposure concentration of a toxic substance lethal to 50% of a test population. See median lethal concentration.

*LD50:* This abbreviation is used for the dose of a toxic substance lethal to 50% of a test population. See median lethal dose.

*lower explosive limit (LEL):* The lower limit of flammability of a gas or vapour at normal ambient temperatures expressed as percentage of the gas or vapour in air by volume. This limit is assumed constant for temperatures up to 130°C.

## M

*mass mean diameter:* The diameter of a particle with a mass equal to the mean mass of all the particles in the population (IAEA, 1978).

*maximum allowable concentration (MAC):* Exposure concentration not to be exceeded under any circumstances.

*maximum residue limit:* The maximum concentration of a pesticide residue resulting from the use of a pesticide according to good agricultural practice directly or indirectly for the production and/or protection of the commodity for which the limit is recommended. The maximum residue limit should be legally recognized. It is expressed in milligrams of the residue per kilogram of the commodity (WHO, 1976).

*median lethal concentration:* Statistically derived concentration of a chemical in water solution that can be expected to cause death in 50% of given population of organism under defined set of experimental conditions.

*median lethal dose:* Statistically derived single dose of a chemical that can be expected to cause death in 50% of given population of organism

192 *Glossary*

under defined set of experimental conditions (for example oral administration, rat).

*metabolic transformation (synonym: biotransformation):* The chemical transformation of substances that takes place within an organism (WHO, 1979).

*metabolism:* In general, the sum total of all physical and chemical processes that take place within an organism; in a narrower sense, the physical and chemical changes that take place in a given chemical sub- stance within an organism. It includes the uptake and distribution within the body of chemical compounds, the changes (biotransformations) undergone by such substances, and the elimination of the compounds and of their metabolites (WHO, 1979).

*metabolite:* A substance resulting from chemical transformation in an organism (WHO, 1979).

*morbidity:* Any departure, subjective or objective, from a state of physiological or psychological well-being. In this sense, sickness, illness, and morbid condition are similarly defined and synonymous (Last, 1988). The WHO Expert Committee on Health Statistics noted in its Sixth report (1959) that morbidity could be measured in terms of three units: (i) persons who were ill; (ii) the illnesses (periods or spells of illnesses) that these persons experienced; and (iii) the duration (days, weeks, etc.) of these illnesses (Last, 1988).

*morbidity survey:* A method for the estimation of the prevalence and/or incidence of disease or diseases in a population. A morbidity survey is usually designed simply to ascertain the facts as to disease distribution, and not to test a hypothesis (Last, 1988).

*mortality rate:* See death rate.

*multigeneration study:* Toxicity test in which at least 3 generations of the test organism are exposed to the substance being assessed. Exposure is usually continuous.

*mutagenicity:* The property of a physical, chemical, or biological agent to induce mutations in living tissue (WHO, 1979).

*mutagen:* An agent that induces mutation (WHO, 1979).

# Glossary 193

*mutation:* Any heritable change in genetic material. This may be a chemical transformation of an individual gene (a gene or point mutation), which alters its function. On the other hand, this change may involve a rearrangement, or a gain or loss of part of a chromosome, which may be microscopically visible. This is designated a chromosomal mutation (WHO, 1979).

## N

*natural occurrence:* The occurrence in nature of a compound, when there are no man-made sources of the compound. The contamination of nature by some compounds may be so widespread that it is virtually impossible at the present time to get access to biota with a natural level and only ``normal" levels can be measured, i.e., the levels that are usually prevalent at places where there is no obvious local contamination (WHO, 1979).

*necrosis:* Mass death of areas of tissues surrounded by otherwise healthy tissue.

*neoplasm:* Any formation of tissue assiciated with disease such as tumour. See malignant, tumour.

*no-observed-adverse-effect-level (NOEL):* The greatest concentration or amount of a chemical, found by experiment or observation, that causes no detectable adverse alteration of morphology, functional capacity, growth, development, or life span of the target (WHO, 1979).

## O

*objective environment:* The actual physical, chemical, and social environment as described by objective measurements, such as noise levels in decibels and concentrations of air pollutants (WHO, 1979).

*occupational environment:* The environment at a work place (WHO, 1979).

*occupational hygiene:* The applied science concerned with recognition, evaluation and control of chemical, physical or biological factors arising from the workplace and which may affect the well-being of those at work or in the community.

*odds:* The ratio of the probability of occurrence of an event to that of non-occurrence, or the ratio of the probability that something is so, to the probability that it is not so (from Last, 1983).

*odds ratio:* The ratio of two odds. The term ``odds" is defined differently according to the situation under discussion. Consider the following notation for the distribution of a binary exposure and a disease in a population or a sample.

|  | Exposed | Unexposed |
|---|---|---|
| Disease | a | b |
| No disease | c | d |

The odds ratio (cross-product ratio) is ad/bc.

The disease-odds (rate-odds) ratio for a cohort or cross-section is the ratio of the odds in favour of disease among the exposed (a/c) to the odds in favour of disease among the unexposed (b/d). The prevalence-odds ratio refers to an odds ratio derived cross- sectionally, as, for example, an odds ratio derived from studies of prevalent (rather than incident) cases. The risk-odds ratio is the ratio of the odds in favour of getting disease, if exposed, to the odds in favour of getting disease if not exposed. The odds ratio derived from a cohort study is an estimate of this (Last, 1983).

*oxidizing:* This adjective applied to a chemical is a substance which gives off oxygen to another substance. Oxidizing chamicals may increase ans sustain fires. For example, chemicals belonging to following groups may act as oxidizers: bromates, chlorates, chromates, dichromates, iodates, nitrates, oxides, perborate, perbromates, perchlorates, periodates, permanganates and peroxodes.

**P**

*persistence:* When applied to a chemical this has a meaning of ability to remain unchanged in the environment.

## Glossary 195

*pesticides:* This is a descriptor applied to chemicals used to kill pests and minimize their impact in agriculture, health and other human interests. Pesticides are often classified according to the organisms which they are used to control, for example as fungicides, herbicides, insecticides, molluscicides, nematicides, rodenticides, etc.

*pesticide residue:* A pesticide residue is any substance or mixture of substances in food for man or animals resulting from the use of a pesticide and includes any specified derivatives, such as degradation and conversion products, metabolites, reaction products, and impurities that are considered to be of toxicological significance (WHO, 1976).

*pH:* A mean to express and to compare the acidity and alkalinity of a solution. It is expressed in a scale from 0 to 14. The solution of pH 7 is neutral; if the pH is lower than 7 the solution is acidic; if the pH is higher than 7 the solution is alkaline (basic).

*pollutant:* Any undesirable solid, liquid, or gaseous matter in a gaseous, liquid, or solid medium (ISO, 1977). For the meaning of ``undesirable" in air pollution contexts, see pollution. A primary pollutant is a pollutant emitted into the atmosphere from an identifiable source. A secondary pollutant is a pollutant formed by chemical reaction in the atmosphere (WHO, 1980).

*pollution:* The introduction of pollutants into a solid, liquid, or gaseous medium, the presence of pollutants in a solid, liquid, or gaseous medium, or any undesirable modification of the composition of a solid, liquid, or gaseous medium (ISO, 1979). For air pollution, an undesirable modification is one that has injurious or deleterious effects.

*population (statistics):* The totality of items under consideration. Every clearly defined part of a population is called a ``subpopulation". In the case of a random variable, the probability distribution is considered as defining the population of that variable (ISO, 1977). The term Population Segment is sometimes used as a synonym for subpopulation.

*population at risk:* The number of people who can develop the adverse health effect under study and who are potentially exposed to the risk factor of interest. For example, all people in a population who have not developed immunity to an infectious disease are at risk of developing

196 *Glossary*

the disease, if they are exposed. Similarly, people already having chronic disease are excluded from the population at risk in studies of the incidence of the disease (WHO, 1979).

*population critical concentration (PCC):* The concentration of a chemical in the critical organ (toxicology) at which a specified percentage of the exposed population has reached their individual critical organ concentrations. The percentage indicated by PCC-10 for 10%, PCC- 50 for 50% etc. (similar to the use of the term LC50) (Kjellström et al., 1984).

*potential daily intake:* The potential daily intake of a pesticide is the theoretical intake calculated on the basis of the maximum residue limits and/or extraneous residue limits and the per caput consumption of the relevant food commodities per day. The same concept applies to food additive intakes (Vettorazzi, 1980).

*potentiation:* The joint action of two or more chemicals on an organism is more than additive (WHO, 1978a).

*ppb:* parts per billion

*ppm:* parts per million

*precision:* The closeness of agreement between the results obtained by applying the experimental procedure several times under prescribed conditions (ISO, 1977).

*prevalence:* The number of instances of a given disease or other condition in a given population at a designated time; sometimes used to mean prevalence rate. When used without qualification, the term usually refers to the situation at a specified point in time (point prevalence).

*prevalence, annual(an occasionally used index):* The total number of persons with the disease or attribute at any time during a year. It includes cases of the disease arising before but extending into or through the year as well as those having their inception during the year.

*prevalence, lifetime:* The total number of persons known to have had the disease or attribute for at least part of their life.

*prevalence, period:* The total number of persons known to have had the disease or attribute at any time during a specified period.

# Glossary 197

*prevalence, point:* The number of persons with a disease or an attribute at a specified point in time (Last, 1988).

*prevalence rate (ratio):* The total number of individuals who have an attribute or disease at a particular time (or during a particular period) divided by the population at risk of having the attribute or disease at this point in time or midway through the period. A problem may arise with calculating period prevalence rates because of the difficulty of defining the most appropriate denominator (Last, 1988).

*primary pollutant:* See pollutant.

*primary protection standard:* An accepted maximum level of a pollutant (or its indicator) in the target, or some part thereof, or an accepted maximum intake of a pollutant or nuisance into the target under specified circumstances (UN, 1972).

*proportionate mortality rate, ratio (PMR):* Number of deaths from a given cause in a specified time period, per 100 or 1000 total deaths in the same time period. Can give rise to misleading conclusions if used to compare mortality experience of populations with different distributions of causes of death (Last, 1988).

*public health impact assessment:* Application of risk assessment procedures to a specific target population. The size of the populations needs to be known. The end product is a quantitative statement about the number of people affected in the specific target populations (WHO, 1988).

## R

*radioactive half-life:* (i) For a single radioactive decay process, the time required for the activity to decrease to half its value by that process (ISO, 1972); (ii) the time taken for the activity of an amount of radioactive nuclide to fall to half its initial value (ICRU, 1980).

*rate:* A measure of the frequency of a phenomenon. An expression of the frequency with which an event occurs in a defined population (from Last, 1988).

*rate difference (RD):* The absolute difference between two rates, for example, the difference in incidence rate between a population group

198 *Glossary*

exposed to a causal factor and a population group not exposed to the factor:

$$RD = Ie - Iu$$

where Ie= incidence rate among exposed, and Iu = incidence rate among unexposed. In comparisons of exposed and unexposed groups, the term excess rate may be used as a synonym for rate difference (Last, 1988).

*rate ratio (RR):* The ratio of two rates. The term is used in epidemiologic research with a precise meaning, i.e., the ratio of the rate in the exposed population to the rate in the unexposed population:

$$RR = \frac{Ie}{Iu}$$

where Ie is the incidence rate among exposed and Iu is the incidence rate among unexposed (Last, 1988).

reference population: The standard against which a population that is being studied can be compared (Last, 1988).

*relative risk:* (i) The ratio of the risk of disease or death among the exposed to the risk among the unexposed; this usage is synonymous with risk ratio; (ii) alternatively, the ratio of the cumulative incidence rate in the exposed to the cumulative incidence rate in the unexposed, i.e., the cumulative incidence ratio, and (iii) the term ``relative risk'' has also been used synonymously with ``odds ratio'' and, in some biostatistical articles, has been used for the ratio of forces of morbidity. The use of the term ``relative risk'' for several different quantities arises from the fact that for ``rare'' diseases (e.g., most cancers) all the quantities approximate one another. For common occurrences (e.g., neonatal mortality in infants under 1500g birth weight), the approximations do not hold (Last, 1988).

*renal elimination:* Excretion of a substance through the kidneys

*reproductive effects:* The adverse effects of a chemical on any aspects of reproduction in an organism (WHO, 1979).

# Glossary 199

*response:* The proportion of an exposed population with an effect or the proportion of a group of individuals that demonstrate a defined effect in a given time (e.g., death) (WHO, 1979).

*risk:* The probability that an event will occur, e.g., that an individual will become ill or die within a stated period of time or age. Also, a nontechnical term encompassing a variety of measures of the probability of a (generally) unfavourable outcome (Last, 1988). Risk should not be confused with the term "hazard". Risk is most correctly applied to predicted or actual frequency of occurrence of an adverse effect of a chemical or other hazard.

*risk assessment:* A combination of hazard identification, quantification of risk resulting from a specific use or occurrence of a chemical, taking into account the possible harmful effects on individual people or society of using the chemical in the amount and manner proposed and all the possible routes of exposure. Quantification ideally requires the establishment of dose-effect and dose-response relationships in likely target individuals and populations. Compare "risk evaluation".

*risk assessment management process:* A global term for the whole activity from hazard identification to risk management (WHO, 1988).

*risk characterization:* The outcome of hazards identification and risk estimation applied to a specific use or occurrence of an environmental health hazard (e.g., a chemical compound). The assessment requires quantitative data on the human exposure in the specific situation. The end product is a quantitative statement about the proportion of affected people in a target population (WHO, 1988).

*risk estimation:* The quantification of dose-effect and dose-response relationships for a given environmental agent, showing the probability and nature of the health effects of exposure to the agent (WHO, 1988).

*risk evaluation:* Risk evaluation involves the establishment of qualitative or quantitative relationship between risks and benefits, involving the complex process of determining the significance of identified hazards and estimated risks to those organisms or people concerned with or affected by them.

200 *Glossary*

*risk management:* The managerial, decision-making and control process to deal with those environmental agents for which risk evaluation has indicated that the risk is too high (WHO, 1988).

*risk monitoring:* The process of following up decisions and actions within risk management in order to check whether the aims of reduced exposure and risk are achieved (WHO, 1988).

*rodenticide:* A chemical used to kill rodents (rats).

## S

*safety factors:* A factor applied to the no-observed-effect level to derive acceptable daily intake (ADI) (the no-observed-effect level is divided by the safety factor to calculate the ADI). The value of the safety factor depends on the nature of the toxic effect, the size and type of population to be protected, and the quality of the toxicological information available (WHO, 1987).

*screening:* The presumptive identification of unrecognized disease or defect by the application of tests, examinations, or other procedures which can be applied rapidly. A screening test is not intended to be diagnostic. Persons with positive or suspicious findings must be referred to their physicians for diagnosis and necessary treatment.

*SDS:* Security data sheets

*secondary pollutant:* See pollutant.

*sensitivity and specificity:* Sensitivity is the proportion of truly diseased persons in the screened population who are identified as diseased by the screening test. Sensitivity is a measure of the probability of correctly diagnosing a case, or the probability that any given case will be identified by the test (synonym: true positive rate).

*sensitization:* This term is applied to the exposure to a substance (allergen) which provokes a response in the immune system such that disease symptoms will ensue on subsequent encounters with the same substance. See hypersensitivity, immune system.

*short term exposure limit (STEL):* According to American Conference of Government Industrial Hygienists, this is the time-weighted average

# Glossary 201

(TWA) airborne concentration to which workers may be exposed for periods up to 15 minutes, which no more than 4 such excursions per day and at least 60 minutes between them. See time-weighted average.

*standardized mortality (morbidity) ratio (SMR):* The ratio of the number of events observed in the study group or population to the number of deaths expected if the study population had the same specific rates as the standard population, multiplied by 100 (Last, 1988).

*stochastic effect:* Effect for which the probability of occurrence depends on the absorbed dose. Hereditary effects and cancer induced by radiation are considered to be stochastic effects (ICRP, 1977). The term ``stochastic'' indicates that the occurrence of effects so named would be random. This means that, even for an individual, there is no threshold of dose below which the effect will not appear, but the chance of experiencing the effect increases with increasing dose (WHO, 1979).

*subacute toxicity test:* An animal experiment serving to study the effects produced by the test material when administered in repeated doses (or continuously in food, drinking water) over a period of up to about 90 days (WHO, 1979).

*subjective environment:* The environment as it is perceived by persons living in it, e.g., eye irritation caused by air pollution, or pleasure arising from good housing conditions (WHO, 1979).

*surveillance:* Ongoing scrutiny, generally using methods distinguished by their practicability, uniformity, and frequently their rapidity, rather than by complete accuracy. Its main purpose is to detect changes in trend or distribution in order to initiate investigative or control measures (Last, 1988).

*synergistic effect:* A synergistic effect is any effect of two chemicals acting together which is greater than a simple sum of their effects when acting alone.

*systemic toxicity:* This term is applied when a substance affects a system in the organism other than and often distant from the site of application or exposure.

## T

*target (biological):* Any organism, organ, tissue, or cell that is subject to the action of a pollutant or other chemical, physical, or biological agent (WHO, 1979).

*target (of environmental pollution):* A human being or any organism, organ, tissue, cell, resource, or any constituent of the environment, living or not, that is subject to the activity of a pollutant or other chemical or physical activity or other agent (WHO, 1979).

*target organ(s):* Organ(s) in which the toxic injury manifests itself in terms of dysfunction or overt disease (WHO, 1979).

*target population:* (i) The collection of individuals, items, measurements, etc., about which we want to make inferences. The term is sometimes used to indicate the population from which a sample is drawn and sometimes to denote any ``reference'' population about which inferences are required; (ii) The group of persons for whom an intervention is planned (Last, 1988).

*temporary acceptable daily intake:* Used when data are sufficient to conclude that use of the substance is safe over the relatively short period of time required to generate and evaluate further safety data, but are insufficient to conclude that use of the substance is safe over a lifetime. A higher-than-normal safety factor is used when establishing a temporary ADI and an expiration date is established by which time appropriate data to resolve the safety issue should be available (WHO, 1987).

*temporary maximum residue limit:* A temporary maximum residue limit is a maximum residue limit established for a specified, limited period when (i) only a temporary or conditional acceptable daily intake has been established for the pesticide concerned, or (ii) although an acceptable daily intake has been established, the residue data are inadequate for firm maximum residue recommendations (WHO, 1976).

*teratogen:* This is the descriptor applied to any substance that can cause non-heritable birth defects.

# Glossary 203

*teratogenicity:* The property (or potential) to produce structural malformations or defects in an embryo or fetus (WHO, 1987).

*threshold limit value (TLV):* This is a quideline value defined by the by the American Conference of Government Industrial Hygienists to establish the airborne concentration of a potentially toxic substance to which it is believed that healthy working adults may be exposed safely through a 40 hour working week and a full working life. This concentration is measured as a time-weighted average concentration. They are developed only as a quidelines to assist in the control of health hazards and are not developed for use as legal standards.

*time-weighted average (TWA) exposure:* This is a regulatory value defining the concentration of a substance to which a person is exposed in ambient air divided by the total time of observation. For occupational exposure a working shift of eight hours is commonly used as the averaging time.

*tolerance:* Tolerance is the ability to experience exposure to potentially harmful amounts of a substance without showing an adverse effect. An adaptive state characterized by diminished responses to the same dose of a chemical (WHO, 1979).

*toxicity:* The toxicity of a substance is the capacity to cause injury to a living organism (WHO, 1978a). A highly toxic substance will cause damage to an organism if administered in very small amounts and a substance of low toxicity will not produce an effect unless the amount is very large. However, toxicity cannot be defined in quantitative terms without reference to the quantity of substance administered or absorbed, the way in which this quantity is administered (e.g., inhalation, ingestion, injection) and distributed in time (e.g., single or repeated doses), the type and severity of injury, and the time needed to produce the injury (WHO, 1979).

- *acute toxicity*: Adverse effects occurring within a short time of administration of a single dose of a chemical, or immediately following short or continuous exposure, or multiple doses over 24 hours or less.

- *subacute toxicity*: Adverse effects occurring as a result of repeated daily dosing of a chemical, or exposure of the chemical, for part of an

204 *Glossary*

organism's lifespan (usually not exceeding 10%). With experimental animals, the period of exposure may range from a few days to 6 months. - *chronic toxicity*: Adverse effects occurring as a result of repeated dosing with a chemical on a daily basis, or exposure of the chemical, for large part of an organism's lifespan (usually more than 50%). With experimental animals, this usually means a period of exposure of more than 3 months. Chronic exposure studies over 2 years using rats or mice are used to assess the carcinogenic potential of chemicals.

*toxicokinetics:* A term with the same meaning as chemobiokinetics for substances not used as drugs (WHO, 1979).

*toxicometry:* A combination of investigation methods and techniques for making a quantitative assessment of toxicity and hazards of poisons (UNEP/IRPTC, 1982).

*tumour: (neoplasm):* This term describes any growth of tissue forming an abnormal mass. Cells of a benign tumour will not spread and will not cause cancer. Cells of a malignant tumour can spread through the body and cause cancer.

## U

*units of measurement:* The base units of the SI system are: metre (m), kilogram (kg), second (s), ampere (A), kelvin (K), candela (cd), and mole (mol) (BIPM, 1979).

*uptake (synonym: absorption):* The entry of a chemical substance into the body, into a cell, or into the body fluids by passage through a membrane or by other means (WHO, 1979).

## X

*Xenobiotic:* A xenobiotic is a chemical which is not natural component of the organism exposed to it. Synonyms: drug, foreign substance or compound.

# AUTHOR'S CONTACT INFORMATION

***Maria Pia Gatto, PhD***
Chemical Researcher
Inail, Department of Occupational and Environmental Medicine,
Epidemiology and Hygiene (DiMEILA), Italy
Email: mp.gatto@gmail.com

# INDEX

## A

accident prevention, 180
As Low As Reasonably Acceptable
(ALARA), 93, 94
asbestos, 134
autoclave, 12, 33, 34, 35, 73, 74, 77, 80, 82, 85

## B

biological risk, 7, 69, 70, 71, 78, 112
biomaterials, 166, 167
biosafety, 70, 71, 72, 73, 74, 76, 81, 83, 107, 116, 121
biosafety level, 71, 72, 73, 74, 75, 76, 77, 81, 82, 84, 85, 86
breathing, 24, 45, 53, 136, 156, 157, 161, 183
burn(s), 8, 13, 34, 35, 36, 37, 52, 53, 104, 155, 156, 161

## C

cancer, 48, 53, 183, 201, 204
car accidents, 159

carbon dioxide ($CO_2$), 24, 27, 60, 61, 149
carbon tetrachloride, 61
carcinogenesis, 183
cardiac arrest, 13, 160
catalyst, 111, 186
centrifuges, 9, 12, 33, 166, 167, 171
chemical degradation, 138
chemical pneumonia, 48
chemical properties, 58
chemical risk, 7, 43
chimneys, 18, 29
chlorinated hydrocarbons, 60
cleaning, 29, 32, 36, 58, 81, 132, 133, 135, 162, 184
clothing, ix, 24, 40, 75, 77, 82, 84, 123, 133, 134, 135, 136, 137, 146, 147, 148, 149, 154, 155, 156, 161, 162, 169, 170
combustion, 28, 47, 57, 141, 143
communication, 46, 138
compatibility, 65, 122, 151
compliance, 23, 36, 43, 64, 90, 95
compounds, 8, 60, 61, 66, 120, 186, 192, 193
conditioning, 7, 29, 30, 31, 85
construction, 23, 136, 190
containment level, 71, 79
contaminant, 31, 73, 111, 112, 184

# Index

contamination, 74, 78, 79, 80, 95, 112, 113, 129, 136, 145, 147, 193
control measures, 103, 119, 121, 201
controlled exposure, 5

## D

danger, 3, 6, 11, 12, 14, 18, 23, 26, 33, 38, 44, 111, 127, 141, 160, 187, 188
death rate, 185, 189, 192
deaths, 197, 201
decomposition temperature, 58
decontamination, 73, 74, 76, 77, 80, 82, 84, 85, 136
Department of Energy, 91
Department of Health and Human Services, 87
dose limit, 90, 94

## E

electric shock, 13, 160
emergency, 5, 11, 18, 30, 40, 58, 82, 84, 109, 134, 135, 136, 141, 146, 147, 148, 149, 155, 160, 161, 168, 180
emergency response, 134, 135, 136
engineering, 2, 49, 58, 91, 107, 108, 129, 133
environment, 2, 3, 7, 8, 10, 28, 47, 50, 54, 55, 62, 70, 72, 76, 78, 79, 80, 108, 112, 139, 173, 180, 182, 185, 187, 190, 193, 194, 201, 202
environmental hazards, 44
equipment, ix, 5, 7, 9, 12, 13, 19, 24, 27, 31, 33, 39, 40, 57, 58, 70, 71, 73, 74, 75, 76, 77, 79, 81, 82, 91, 95, 103, 107, 121, 133, 134, 136, 139, 142, 146, 147, 148, 149, 150, 151, 161, 162, 165, 166, 167, 169, 171, 172, 173, 184, 187
ethylene glycol, 60, 61
evacuation, 20, 148, 149, 162, 180

exposure, 3, 5, 13, 14, 15, 30, 31, 33, 38, 39, 44, 45, 46, 48, 49, 53, 54, 57, 58, 67, 70, 71, 79, 89, 91, 92, 93, 94, 104, 107, 108, 111, 112, 119, 122, 128, 129, 133, 134, 135, 137, 161, 180, 181, 182, 183, 184, 185, 186, 187, 188, 190, 191, 194, 199, 200, 201,203, 204
exposure limit, 5, 45, 46, 49, 58, 180, 187, 200
extinguishers, 11, 24, 39, 143, 144, 145, 147, 148, 149

## F

fire classes, 142, 143
fire hazard, 36, 105
fire resistance, 30, 62
fire safety, 7, 141, 149, 150
fires, 141, 142, 143, 144, 145, 146, 148, 149, 194
first aid, 7, 35, 40, 153, 156, 157, 188
flame, 19, 26, 28, 36, 50, 120, 141
flammability, 27, 58, 191
flue gas, 109
fume hoods, 28, 107, 108, 109, 110, 117, 146, 150

## G

gas cylinders, 17, 20, 24, 25, 26, 63, 64, 150, 166, 167
gas generators, 23, 27
generic risks, 7
glassware, 11, 31, 32, 35, 40, 41, 121
Globally Harmonized System (GHS), 46, 48, 49, 50, 56, 57, 59, 67
gravity cycle, 35

## Index

### H

hazardous materials, 31, 66, 148, 150
hazardous substance, 4, 30, 39, 40, 45, 47, 62, 120
hazardous substances, 4, 30, 39, 40, 47, 62, 120
hazardous waste, 17, 40, 95, 134, 135, 136
hazards, 12, 33, 44, 47, 50, 56, 57, 59, 67, 71, 78, 105, 107, 111, 123, 127, 128, 129, 130, 133, 134, 136, 150, 162, 165, 166, 167, 184, 199, 203, 204
health hazards, 44, 48, 203
hydrogen, 26, 27, 28, 47, 60, 61, 158, 159
hydrogen atoms, 47
hydrogen peroxide, 47, 60, 61, 158, 159
hydrogen sulfide, 60

### I

intoxications, 153, 154
ionizing radiation, 4, 89, 90, 91, 93, 94, 103, 179, 190

### L

laboratory, ix, x, 2, 4, 5, 7, 8, 9, 11, 12, 13, 14, 15, 16, 17, 19, 23, 26, 27, 28, 29, 30, 31, 32, 33, 34, 35, 36, 37, 39, 40, 41, 43, 49, 62, 64, 65, 67, 70, 71, 72, 73, 74, 75, 76, 77, 79, 80, 81, 82, 83, 84, 85, 86, 87, 89, 92, 95, 96, 97, 106, 107, 108, 109, 110, 112, 113, 117, 126, 127, 141, 146, 148, 149, 150, 151, 152, 157, 165, 166, 167, 168, 169, 171, 172, 173, 174, 175, 176, 179
liquid cycle, 35
liquid nitrogen, 8, 37, 38, 166, 167

### M

materials, 8, 9, 11, 19, 23, 24, 28, 36, 37, 40, 45, 58, 60, 61, 62, 63, 64, 65, 69, 73, 75, 78, 82, 83, 90, 91, 93, 96, 97, 108, 109, 112, 121, 122, 124, 125, 126, 127, 129, 136, 137, 139, 141, 146, 148, 151, 162, 165, 166, 167, 169, 170, 171, 172, 180, 182, 188
microclimate, 4, 7
microorganisms, 69, 72, 78, 85, 121
mittens, 34

### N

National Institute for Occupational Safety and Health, 140
National Institutes of Health, 72, 87
National Research Council, 67
nitrogen, 8, 26, 27, 37, 38, 166, 167
noble gases, 24, 26
noise, 4, 9, 119, 138, 165, 166, 167, 193
Nuclear Regulatory Commission, 106

### O

operations, 3, 9, 10, 31, 35, 36, 39, 40, 79, 80, 121, 126, 129, 134, 135, 139, 144
OSHA, 18, 31, 43, 45, 46, 56, 67, 127, 140, 143, 146, 168, 176
oxygen, 26, 27, 47, 61, 130, 141, 144, 145, 160, 185, 194

### P

Permissible Exposure Limit (PEL), 45
personal protective equipment (PPE), 9, 18, 33, 35, 40, 58, 70, 74, 75, 76, 77, 91, 107, 119, 120, 121, 122, 133, 138, 139, 140, 146, 162

210          *Index*

physical hazards, 44, 47, 107, 127

poison control centers, 50

pollutant, 59, 180, 184, 186, 195, 197, 200, 202

protection, ix, 2, 6, 9, 22, 28, 31, 33, 38, 49, 58, 73, 74, 76, 79, 90, 91, 108, 113, 119, 121, 122, 127, 128, 129, 130, 131, 133, 134, 135, 136, 142, 151, 170, 184, 187, 191, 197

protective clothing, 24, 123, 133, 134, 135, 136, 137, 140, 161, 162, 169, 170

## Q

Q fever, 76

## R

radioactive waste, 92, 98

radiological safety, 7, 89, 90

respiratory protection, 129, 140

risk assessment, 3, 4, 5, 6, 17, 22, 45, 70, 71, 72, 121, 197, 199

## S

safety, ix, x, 5, 8, 11, 17, 18, 20, 23, 24, 26, 27, 28, 29, 33, 34, 39, 40, 41, 46, 49, 55, 56, 62, 70, 71, 74, 75, 76, 77, 80, 83, 85, 89, 90, 93, 96, 97, 107, 108, 112, 113, 121, 122, 128, 129, 143, 147, 149, 150, 151, 152, 162, 165, 166, 167, 168, 169, 173, 179, 200, 202

Safety Data Sheets (SDS), 49, 50, 56, 59, 65, 98, 108, 161, 162, 200

safety signs, 18

Short-Term Exposure Limit (TLV-STEL), 46

solvents, 62, 65, 66, 82, 126, 141, 142, 146, 149

spill, 148, 161, 162, 163, 164, 165

storage, 17, 23, 26, 37, 49, 58, 62, 63, 64, 82, 90, 91, 107, 146, 149, 150, 151, 165, 167

## T

Threshold Limit Value (TLV), 45, 46, 108, 161, 203

Time-Weighted Average TLV-TWA (TLV-TWA), 46

Total Effective Dose Equivalent (TEDE), 93, 94

toxic effect, 200

toxic gases, 17, 24, 30

toxic substances, 155

toxicity, 48, 58, 69, 120, 141, 154, 155, 179, 180, 181, 183, 201, 203, 204

toxicology, 179, 184, 196

training, 5, 30, 49, 79, 83, 90, 91, 136, 140, 141

## U

United Nations, 46, 48, 67

United States (USA), 16, , 81, 86, 168

## V

ventilation, 8, 18, 24, 30, 31, 62, 76, 82, 83, 84, 85, 107, 112, 129, 149

video terminals, 14

## W

water, 11, 13, 19, 25, 28, 38, 44, 47, 52, 54, 58, 60, 61, 69, 82, 84, 85, 123, 138, 143, 149, 151, 154, 155, 156, 157, 158, 161, 162, 164, 165, 182, 186, 190, 191

work environment, 14, 112

## Index

workers, ix, 5, 6, 9, 18, 43, 46, 49, 72, 76, 79, 93, 107, 112, 141, 146, 147, 165, 166, 180, 189, 201

working conditions, 14, 184

workplace, 3, 5, 9, 18, 34, 45, 49, 56, 166, 167, 184, 194

World Health Organization, 67, 71, 72, 81, 87

World Health Organization (WHO), 71, 81, 87, 179, 181, 182, 183, 184, 185, 186, 187, 188, 189, 190, 191, 192, 193, 195, 196, 197, 198, 199, 200, 201, 202, 203, 204

wounds, 157, 158

## Z

zinc, 61

zirconium, 142

# Related Nova Publications

## ADVANCES IN HEALTH AND NATURAL SCIENCES

**EDITORS:** Burcu Yuksel and Mustafa Sencer Karagul

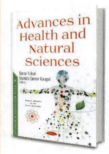

**SERIES:** Public Health in the 21st Century

**BOOK DESCRIPTION:** This book, organized into nine chapters, features scientists from around the globe contributing diverse topics in mostly natural, biological and health sciences. The edited book aims at highlighting the state of the art research and recent findings in of agricultural, environmental, biological, marine and medical sciences and biotechnology, and bridging theoretical research with current applications.

**HARDCOVER ISBN:** 978-1-53614-639-4
**RETAIL PRICE:** $230

## HEALTH AND FREEDOM IN THE BALANCE: EXPLORING THE TENSIONS AMONG PUBLIC HEALTH, INDIVIDUAL LIBERTY, AND GOVERNMENTAL AUTHORITY

**EDITORS:** M. Girard Dorsey and Rosemary M. Caron

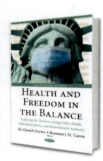

**SERIES:** Public Health in the 21st Century

**BOOK DESCRIPTION:** The clash between individual liberties and the protection of the greater population is an ongoing conflict between core principles held dear by Americans for centuries. One of the nexus points occurs in the application of public health measures by governmental authorities to defeat deadly germs, perhaps on an epidemic scale, in ways that can erode individual decisions about healthcare, privacy, bodily integrity, and personal liberty in the name of the greater good of community health.

**HARDCOVER ISBN:** 978-1-53612-201-5
**RETAIL PRICE:** $160

To see complete list of Nova publications, please visit our website at www.novapublishers.com

# Related Nova Publications

## THE SCIENCE OF MEDICAL CANNABIS

**AUTHOR:** David Steven Younger, M.D.

**SERIES:** Public Health in the 21st Century

**BOOK DESCRIPTION:** The cultural, scientific and legislative divide created by vigorous debates over the legalization of medical marihuana has given way to a new synergy among community stakeholders across the United States to improve access to medical marijuana for patients with refractory debilitating neurological disorders, cancer, and chronic pain as an alternative to ineffective pharmacotherapy and potentially addictive pain medications.

**SOFTCOVER ISBN:** 978-1-53614-566-3
**RETAIL PRICE:** $95

## AN OUTLINE OF ASBESTOS-RELATED HEALTH EFFECTS

**AUTHOR:** David Y. Zhang, M.D., Ph.D.

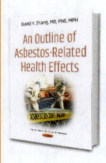

**SERIES:** Public Health in the 21st Century

**BOOK DESCRIPTION:** This book serves as an easy to read, quick reference presented in a bulleted format that allows readers to quickly and easily review the information. It is a useful resource for occupational medicine specialists, healthcare providers and environmental scientists who are interested in understanding asbestos and managing asbestos-related diseases.

**SOFTCOVER ISBN:** 978-1-53610-961-0
**RETAIL PRICE:** $82

*To see complete list of Nova publications, please visit our website at www.novapublishers.com*